# Ecological Research Monographs

**Series Editor**
Yoh Iwasa, Department of Biology, Kyushu University, Fukuoka, Japan

The book series Ecological Research Monographs publishes refereed volumes on all aspects of ecology, including Animal ecology Population ecology Theoretical ecology Plant ecology Community ecology Statistical ecology Marine ecology Ecosystems Biodiversity Microbial ecology Landscape ecology Conservation Molecular ecology Behavioral ecology Urban ecology Physiological ecology Evolutionary ecology The series comprise books and edited collections by international experts in their fields.

Yukari Suzuki-Ohno
Editor

# Community Science in Ecology

## Case Studies of Public Participation in Ecological Research in Japan

Springer

*Editor*
Yukari Suzuki-Ohno
Graduate School of Life Sciences
Tohoku University
Sendai, Miyagi, Japan

ISSN 2191-0707 ISSN 2191-0715 (electronic)
Ecological Research Monographs
ISBN 978-981-97-0303-6 ISBN 978-981-97-0304-3 (eBook)
https://doi.org/10.1007/978-981-97-0304-3



Cover illustration: Sampling environmental DNA from seawater by volunteers in Japan. (Photo taken on September 13, 2022, by Masahiro Mitsuhata)

This Springer imprint is published by the registered company Springer Nature Singapore Pte Ltd.
The registered company address is: 152 Beach Road, #21-01/04 Gateway East, Singapore 189721, Singapore

# Preface

"Citizen science," also called "community science," has attracted the attention of scientists, which is increasing the demand for books on the subject. In light of this, the book *Community Science in Ecology: Case Studies of Public Participation in Ecological Research in Japan* was written. The aim of writing it was twofold. The first was to spread and encourage the practice of community science programs by ecologists. Ecology in the twenty-first century requires ecological community science programs to understand and mitigate the rapid decline in biodiversity or species extinction. Ecologists need to gain more experience and analytical results from community science programs. The authors of this book are all ecologists and practitioners of community science programs and cover the following topics: the introduction of this book (Chap. 1), the basic knowledge about community science (Chap. 2), ecological community science programs (Chaps. 3–9), science communication (Chap. 10), database and the rights of participants (Chap. 11), decision-making in community science (Chap. 12), alien species extermination (Chap. 13), and wetland restoration with participants (Chap. 14). Lastly, in the Editor's note, I have given my opinion as to the challenges of community science programs in Japan.

The second was to introduce Japanese community science programs in English. There is extensive literature on community science, but most of it is from Europe and the United States. This book focuses on Japanese community science programs in the field of ecology because the literature on these programs in Asia is rarely accessible in the English language. Some chapters in this book are not in the general form of an original article in academic journals because this was unsuitable for describing the programs' details. The procedures and outcomes of ecological community science programs often depend on the target species and site locations, and the details of past programs are not always documented to find generality. I believe that it is important to accumulate detailed records of case studies as this book has done. This book will be valuable for scientists studying and practicing community science around the world. I hope that this book will help ecologists who plan or work on ecological community science programs and benefit their participants.

Sendai, Miyagi, Japan Yukari Suzuki-Ohno

# Contents

# Contributors

**Takuzo Abe** Minamisanriku Nature Center, Minamisanriku, Miyagi, Japan

**Akihiro Dazai** Center for Sustainable Society, Minamisanriku, Miyagi, Japan

**Kazumi Fujikawa** The Kochi Prefectural Makino Botanical Garden, Kochi-shi, Kochi, Japan

**Koichi Goka** National Institute for Environmental Studies, Tsukuba, Ibaraki, Japan

**Yuko Ikkatai** Institute of Human and Social Sciences, Kanazawa University, Kanazawa-shi, Ishikawa, Japan

**Yoko Imai** Department of Urban Engineering, School of Engineering, The University of Tokyo, Bunkyo, Tokyo, Japan

**So Ishida** Osaka Museum of Natural History, Higashisumiyoshi-ku, Osaka, Japan

**Minoru Kasada** Graduate School of Life Sciences, Tohoku University, Sendai, Miyagi, Japan

Field Science Center for Northern Biosphere, Hokkaido University, Horokanai, Japan

**Akihide Kasai** Faculty of Fisheries Sciences, Hokkaido University, Hakodate, Hokkaido, Japan

**Michio Kondoh** Graduate School of Life Sciences, Tohoku University, Sendai, Miyagi, Japan

Advanced Institute for Marine Ecosystem Change (WPI-AIMEC), Tohoku University, Sendai, Miyagi, Japan

**Reiji Masuda** Maizuru Fisheries Research Station, Field Science Education and Research Center, Kyoto University, Kyoto, Japan

**Kazuhiro W. Nakamura** Graduate School of Frontier Sciences, The University of Tokyo, Tokyo, Japan

**Jun Nishihiro** Center for Climate Change Adaptation, National Institute for Environmental Studies, Tsukuba, Ibaraki, Japan

**Takeshi Osawa** Graduate School of Urban Environmental Sciences, Tokyo Metropolitan University, Hachiouji, Tokyo, Japan

**Satoquo Seino** Graduate School of Engineering, Kyushu University, Fukuoka, Japan

**Yukari Suzuki-Ohno** Graduate School of Life Sciences, Tohoku University, Sendai, Miyagi, Japan

**Shota Suzuki** Minamisanriku Nature Center, Minamisanriku, Miyagi, Japan

**Shinichi Takagawa** The Nature Conservation Society of Japan, Chuo-Ku, Tokyo, Japan

**Akifumi S. Tanabe** Graduate School of Life Sciences, Tohoku University, Sendai, Miyagi, Japan

**Jun Yokoyama** Yamagata University, Yamagata-shi, Yamagata, Japan

**Masashi Yoshimura** Okinawa Institute of Science and Technology Graduate University, Okinawa, Japan

# Chapter 1
# Introduction

**Yukari Suzuki-Ohno**

**Abstract** For centuries, volunteers have accumulated data on animals and plants, developing what is now known as "citizen science" or "community science." It is meaningful for ecologists to be involved in community science for the future of ecology. This book "Community Science in Ecology: Case Studies of Public Participation in Ecological Research in Japan" covers case studies of community science programs conducted by ecologists (Chaps. 3, 4, 5, and 6), NPO staff (Chap. 7), and museum curators (Chaps. 8 and 9) in Japan. It also covers a wide range of topics needed to study and practice community science: basic knowledge about community science (Chap. 2), science communication (Chap. 10), database and the rights of participants (Chap. 11), decision-making in community science (Chap. 12), alien species extermination (Chap. 13), and wetland restoration with participants (Chap. 14). To understand why this book includes these chapters, I provide an overview of community science in the field of ecology by referring to these chapters in the Introduction. Introduction consists of "Ecological Community Science," "Ecological Community Science for Ecologists," "Ecological Community Science in Japan," "Science Communication and Data Openness of Ecological Community Science," "Decision-making and Conservation Activities in Ecological Community Science," and "Organization of This Book."

**Keywords** Community science · Citizen science · Science communication · Open data · Decision-making · Conservation · Elimination of alien species

## 1.1 Ecological Community Science

The word "citizen science," which scientists have begun to call "community science," has different definitions depending on its context (Chap. 2 of this book); however, here, I would like to define it simply as scientific activities carried out by

Y. Suzuki-Ohno (✉)
Graduate School of Life Sciences, Tohoku University, Sendai, Japan



Y. Suzuki-Ohno (ed.), *Community Science in Ecology*, Ecological Research Monographs, https://doi.org/10.1007/978-981-97-0304-3_1

volunteers. Community science has been developed by volunteers for centuries (Dickinson et al. 2010). There are various types of community science, such as data classification, annotation, transcription, and collection. Data classification and annotation are popular in a wide range of fields (e.g., astronomy), but data collection is more common in ecology (hereafter, community science in ecology is called "ecological community science"). Volunteers who are not ecologists belonging to universities or research institutes have participated in community science and have collected data by observing birds and catching insects. Traditional recording methods for data collection in ecological community science programs include observations (e.g., Sullivan et al. 2009) (Fig. 1.1) and specimens (e.g., Fontaine et al. 2012) (Fig. 1.2). eBird (https://ebird.org/home) is one of the most famous community science programs, collecting observational data (note: eBird also collects photograph and audio data). eBird has published its dataset, the "eBird Observation Dataset," in the open database Global Biodiversity Information Facility (GBIF, https://www.gbif.org/). The dataset contains 1,060,970,490 bird occurrence records with 100% taxon, 100% year, and 99.9% coordinate matches (as of June 2022) (https://www.gbif.org/dataset/4fa7b334-ce0d-4e88-aaae-2e0c138d049e). The collection of observational data is simple and widely used, but it requires participants to visually identify the species. The collection of specimens allows participants to identify species by anatomy or experts' opinions. Numerous new species have been reported based on specimens collected from volunteers (Fontaine et al. 2012). Specimens are suitable for investigating new or alien species because of the critical importance of species identification (Fig. 1.2, Chap. 3).

**Fig. 1.1** Observation of birds in a woodland area

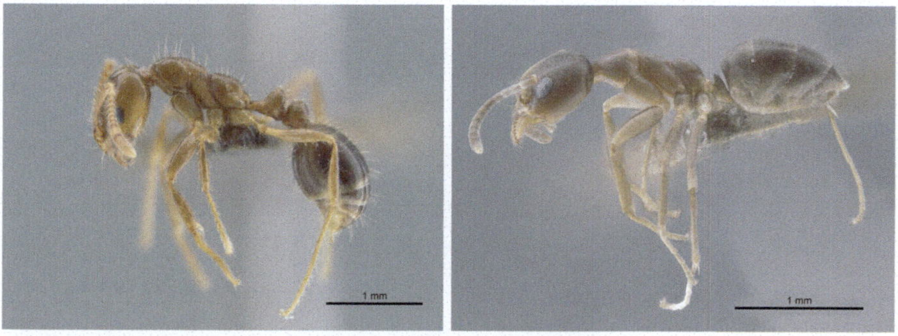


**Fig. 1.2** Alien ant specimens collected in OKEON project (Chap. 3). Left: *Solenopsis invicta*. Right: *Linepithema humile* (Photographer: OKEON)

In the twenty-first century, ecological community science has had the largest impact on the methodology for collecting and classifying data in research on ecology and conservation (Kullenberg and Kasperowski 2016). The prevalence of the Internet and digital tools, such as digital cameras and smartphones, has promoted online ecological community science programs. This allows us to efficiently collect large numbers of high-quality data over a wide area at low costs. The choices of recording methods have increased with the addition of photographs, videos, and audio data (Figs. 1.3 and 1.4). These tools allow the recording of the occurrence data of the target species without capture and the identification of species by experts' or other participants' opinions (e.g., Losey et al. 2012; Silvertown et al. 2015; Saito et al. 2015; Suzuki-Ohno et al. 2017, Chaps. 4 and 5). iNaturalist (https://www.inaturalist.org) and iSpot (https://www.ispotnature.org) are typical online networks that share biodiversity information of photographs. Participants and experts can help identify species in photographs uploaded by other participants on the websites. Photographs are simple and widely used for observation. iNaturalist has published its dataset as "iNaturalist Research-grade Observations" in the open GBIF database. The dataset contains 44,166,226 occurrence records with 99.9% taxon, 100% year, and 99.7% coordinate matches (as of June 2022) (https://www.gbif.org/ja/dataset/50c9509d-22c7-4a22-a47d-8c48425ef4a7). In addition to data collection, data classification and annotation are actively performed using photograph and audio data. For example, animal photographs captured by motion-sensor cameras in the Serengeti National Park of Tanzania were classified by volunteers (Swanson et al. 2015, 2016), and the accuracy of identification of volunteers versus experts was high (98% in Swanson et al. 2016 and 96.6% in Norouzzadeh et al. 2018). Experts and volunteers also identified the bird species through live sounds streamed by automatic fixed cameras in the forest monitoring sites (Saito et al. 2015, Chap. 5).

The latest tool used for community science monitoring is environmental DNA (eDNA) (Biggs et al. 2015; Larson et al. 2017; Tøttrup et al. 2021; Meyer et al. 2021; Miya et al. 2022; Suzuki-Ohno et al. 2023) (Fig. 1.5). The technologies of polymerase chain reaction (PCR), the eDNA databases, and computational theories

**Fig. 1.3** Digital cameras and smartphones. Photograph, video, and audio data recorded by digital cameras and smartphones can be used as biodiversity data

have made it possible to identify species from eDNA and record occurrence data without relying on visible morphological features. In addition, eDNA metabarcoding allows for a comprehensive investigation of biodiversity at a sampling site. For example, 572 taxonomic groups of fish eDNA were collected from 81 coastal sites by volunteers using this method in 2020 and 2021 (Suzuki-Ohno et al. 2023, Chap. 6). This dataset was published in the ANEMONE DB (https://db.anemone.bio/). The details of these recording methods and tools in ecological community science programs are explained in Chaps. 3, 4, 5, and 6 of this book.

## 1.2 Ecological Community Science for Ecologists

As ecological community science has evolved, the types of ecological community science programs have diversified (Pocock et al. 2017; Chap. 2). Some programs are managed solely by only nonexpert volunteers to investigate a specific local site, whereas others are planned by ecologists and investigate a worldwide range. Some programs are conducted solely for fun, whereas others are aimed at collecting accurate data for scientific research. Discussing them at the same level causes

**Fig. 1.4** Fixed camera in Cyberforest (Chap. 5). Video and audio data recorded by fixed cameras can be used as biodiversity data

**Fig. 1.5** Environmental DNA (eDNA) collected by a sampling kit in Earthwatch project (Chap. 6). Left: Filtering seawater to collect eDNA of marine fishes. Right: A kit (syringe and cartridge) to collect eDNA (Photographer: M. Mitsuhata)

misunderstandings and conflicts that prevent the development of ecological community science. When discussing ecological community science, it is important to clarify the following: (1) main members who plan the program (e.g., volunteers, nonprofit organization (NPO)/nongovernmental organization (NGO)/Public Interest Incorporated Association (PIIA) staff, museum curators, or ecologists), (2) the relationship between the main members who plan the program and those who participate or cooperate in the program (e.g., partners, supporters, or participants in other programs), (3) the geographic range covered in the program (e.g., a specific site, region, or nationwide), (4) the method of collecting/classifying/annotating data (e.g., onsite, or online), (5) the method of recording data (e.g., observations, specimens, photographs, videos, audio, or eDNA), and (6) the aim of the program (e.g., an increase of local attachment, environmental education, conservation, or contribution to science).

The type of ecological community science programs that many ecologists are interested in will correspond to data collection by an unspecified number of participants for nationwide scientific research planned by ecologists. This type benefits the ecologists who plan or contribute to them in the following ways. (1) As many volunteers participate in the program, ecologists can obtain more numbers of data. (2) If the program is of the type in which many monitoring sites are set over wide areas or monitoring sites are freely selectable by volunteers, they can obtain data over wide areas. (3) If the program is of the type that monitoring sites are freely selectable by volunteers, the data may include those of places that they would not normally be able to investigate (e.g., private gardens of volunteers and remote islands). (4) The programs can be used for science outreach activities and environmental education through training in data collection, reporting the collected data, and the analytical results of the data to participants. (5) The program could be collaborative if the program is of the type that frequent communication with participants is possible. In this way, participants may improve their monitoring/classifying/analyzing methods. They may provide new data or ideas for research. The last two benefits, (4) and (5), are also common in data classification and annotation by volunteers. The details of these benefits are found in Chaps. 4 and 6 of this book.

Occurrence data collected by participants can be used to study species distribution (e.g., Losey et al. 2012; Miyazaki et al. 2014; Suzuki-Ohno et al. 2017), alien species invasion (e.g., Delaney et al. 2008; Kadoya et al. 2009; Miyazaki et al. 2016; Morii and Nakano 2017), species composition (e.g., Washitani et al. 2020), and phenology (e.g., Henderson et al. 2012). Some scientists have pointed out that the accuracy of data collected by volunteers may sometimes be too low for the use of scientific research. However, several methods to mitigate the effects of spatially biased data (e.g., Phillips et al. 2009; Kramer-Schadt et al. 2013) or to ensure high accuracy of data (e.g., Van Eupen et al. 2022) have been proposed in previous studies. If the data contain images such as photographs or videos, these can be used to study phenotypic variation (e.g., Silvertown et al. 2011), feeding behavior (e.g., Morii and Nakano 2017), and image analysis by deep learning (e.g., Horn et al. 2018; Suzuki-Ohno et al. 2022). Data collected by participants can also be used to study species' functions. In 2022, Cool Earth via dSOIL (https://dsoil.jp/en/) initiated a community science program called "COOL EARTH LAB." In this program, participants receive an experimental culture jar and vials provided by scientists, put the soil into the jar, sample the gas from the jar into the vials multiple times, and send the jar and the vials back to the scientists. The scientists analyze the gas and soil to investigate $NO_2$ concentration, soil bacteria, and their *nosZ* gene. These results will reveal the relationship between soil bacteria and their function of reducing $NO_2$. By devising equipment and manuals in community science programs, data collected by participants can be applied to various studies in ecology.

## 1.3 Ecological Community Science in Japan

Community science programs in Europe and the United States are actively discussed, but it does not mean that there are few community science programs in Asia. There are many ecological community science programs in Asia. In Japan, there has been more than a century of history of ecological community science (Kobori et al. 2016). This is most likely because Japan is a global hotspot for biodiversity and Japanese people have enjoyed feeling seasonal changes from the change of observed organisms (i.e., phenology). There are four distinct seasons, and the species observed vary during each season. Japanese people have enjoyed cherry blossoms and the arrival of swallows in spring, observing and catching insects (e.g., fireflies, beetles, stag beetles, cicadas, and dragonflies) in summer, seeing colored leaves and listening to the singing of insects (e.g., bell crickets, and pine crickets) in autumn, and watching winter birds in winter. In a 1000-year-old book, "*the Makura no Soshi*" (translated as "The Pillow Book" in English) by Sei Shonagon, the author explains how she loved the light of fireflies, the singing of insects, and the sight of wild geese flying (Sei Shonagon n.d.).

Many ecological community science programs in Japan have been managed by NPOs, NGOs, PIIAs, museum curators, scientists, and volunteers. Famous examples of ecological community science programs managed by NPOs and NGOs are "Sea Turtle Survey" of the Sea Turtle Association of Japan (http://www.umigame.org/) and "Monitoring Sites 1000" project of the Nature Conservation Society of Japan (NACS-J) (Chap. 7, https://www.nacsj.or.jp/english/). Sea Turtle Survey investigated the distribution of sea turtle nesting habitats in the Japanese archipelago for the conservation of sea turtles. This survey was initiated by students in an elementary school and a junior high school in the 1950s and continued to increase in participants at various sites, which led to the establishment of the Sea Turtle Association of Japan in 1990. It is still ongoing, which makes it a remarkably long-term survey even by international standards. Monitoring Sites 1000 Project began after establishing NACS-J. The predecessor of NACS-J was formed by local biologists and mountain climbers in 1949, and NACS-J was established to protect nature in 1951 (Fig. 1.6). NACS-J has collaborated with various NPOs, NGOs, volunteer groups, scientists, and administrations. Monitoring Sites 1000 project was initiated by these groups and the Biodiversity Center of Japan in the Ministry of the Environment in 2003. The details of Monitoring Sites 1000 project are described in Chap. 7 of this book. The NPOs and NGOs cooperating with Monitoring Sites 1000 also manage long-term ecological community science programs. One of the NPOs cooperating with Monitoring Sites 1000 is Bird Research (http://www.bird-research.jp/index-e.html). Bird Research conducts several programs that collect long-term monitoring data of birds, such as verandah birdwatching initiated in 2005. They also conduct a program to identify bird species using audio data from forest monitoring sites in Cyberforest (Chap. 5).

Famous examples of ecological community science programs managed by museum curators and scientists include "Field Research of Dandelions in West

**Fig. 1.6** The pamphlet of Nature Conservation Society of Japan (NACS-J)

Japan" (Chap. 8) and "FishPix" (https://fishpix.kahaku.go.jp/fishimage-e/index.html in English, or https://nh.kanagawa-museum.jp/kpmnh-collections/ in Japanese). Field Research of Dandelions in West Japan began in 1975 and has been managed by PIIA Nature Conservation Society of Osaka, museums, and some laboratories of universities. Dandelion specimens in the western regions of Japan were collected every 5 years, and species of dandelions have been identified by museum curators and university faculty in the western regions of Japan (Chap. 8). FishPix is a database of fish photographs built and managed by the Fish Divisions of the Kanagawa Prefectural Museum of Natural History, Odawara, and the National Museum of Nature and Science. FishPix began in 1998 and has collected over 150,000 photographs of fish species (as of June 2022). These photographs were mainly sent by amateur SCUBA divers. The museums also provide a venue for volunteer groups to make presentations on their community science activities (e.g., Osaka Natural History Festival in the Osaka Natural History Center, Kyousei-no-hiroba in the Museum of Nature and Human Activities, Hyogo, and Field Reporter in the Lake Biwa Museum). The details of the ecological community science programs managed by museum curators are explained in Chaps. 8 and 9 of this book.

Famous examples of ecological community science programs managed by scientists and volunteers not covered in PART II were "An identification guide of Japanese moths compiled by everyone" (hereafter it is called "JPMOTH," http://www.jpmoth.org/) and "WEB *sakana-zukan*" (https://zukan.com/fish/). JPMOTH has interesting origins (Jinbo and Suzuki 2006). The predecessor of JPMOTH was an open anonymous Internet forum, and it generated the website of JPMOTH managed by two volunteers and a scientist (Fig. 1.7). Over 10,000 photographs of moth adults and larvae were taken by the participants, and the number of species/subspecies in the photographs was more than 4000 species/subspecies. The website of the identification guide in JPMOTH was started in 2003 and paused in 2010, but the place of communication for moth enthusiasts has moved to the open anonymous Internet forum in JPMOTH and the semi-closed Facebook group "Moth Club." Moth enthusiasts participate in communication by uploading photographs of collected moths and larvae. JPMOTH influenced overseas ecological community science and triggered an ecological community science program called "twmoth" in Taiwan (https://twmoth.tesri.gov.tw/peo/aboutme). The occurrence data of Taiwanase moths collected by twmoth is available (https://www.gbif.org/dataset/e0b8cb67-6667-423d-ab71-08021b6485f3) in GBIF. WEB *sakana-zukan* is a database of fish photographs built and managed by some scientists and volunteers. The WEB *sakana-zukan* began in 2002 and collected over 50,000 photographs of 3648 fish species (as of June 2022). These photographs were often sent by amateur anglers and have been used in scientific studies (Miyazaki et al. 2014, 2016, 2020).

In contrast to the high level of contribution of Japanese volunteers to community science programs in Japan, the contribution of Japanese volunteers to global community science programs may not be significant. According to data from eBird and iNaturalist, the percentage of observers in Japan is relatively low. On the eBird website (https://ebird.org/region/world?yr=all), Japan's records, species, and observers are 1.3%, 5.7%, and 0.5% of the world's total, respectively (as of May 2022). On the iNaturalist website (https://www.inaturalist.org/observations), Japan's records, species, and observers are 0.24%, 4.6%, and 0.38% of the world's total, respectively (as of May 2022). This may be because Japanese volunteers do not know about these global community science programs or have originally

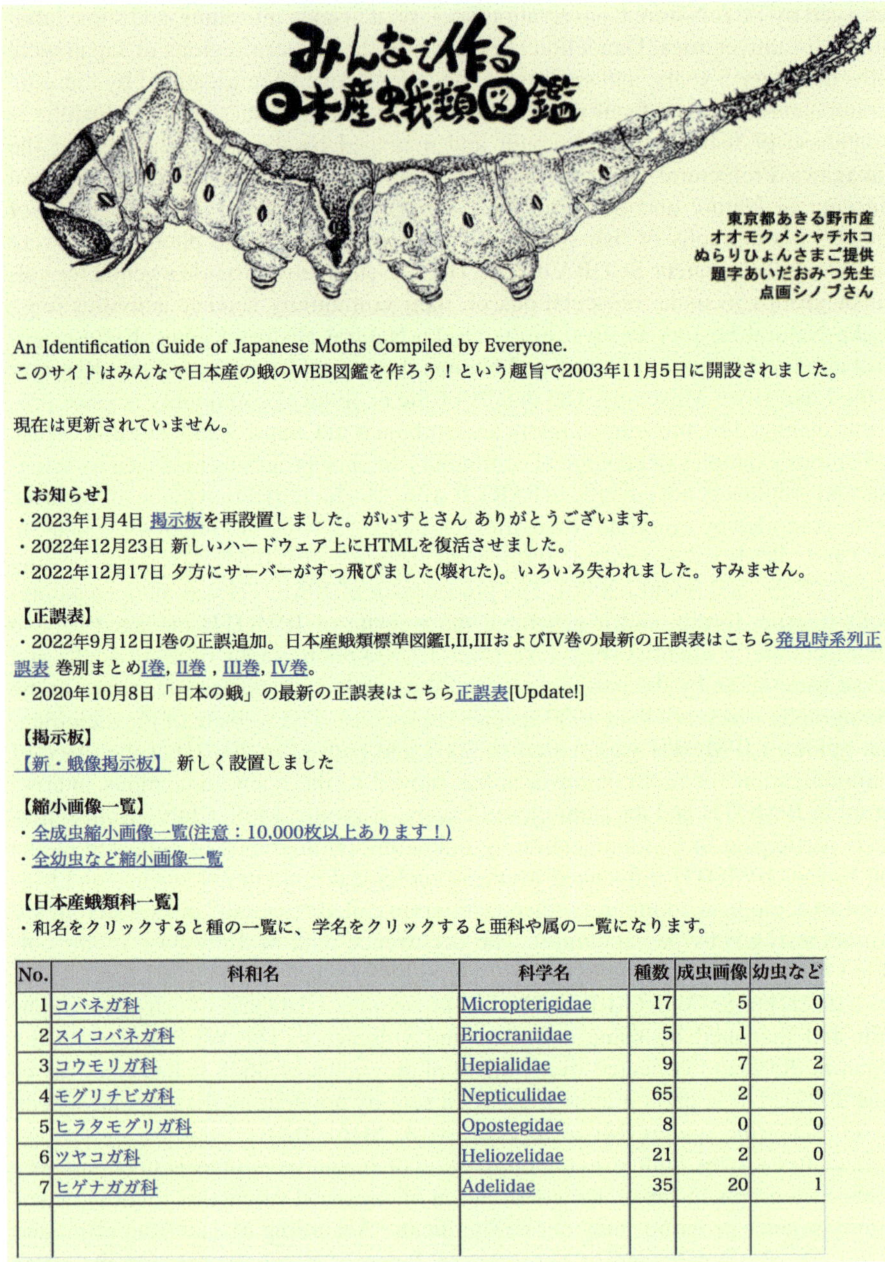
みんなで作る
日本産蛾類図鑑

東京都あきる野市産
オオモクメシャチホコ
ぬらりひょんさまご提供
題字あいだおみつ先生
点画シノブさん

An Identification Guide of Japanese Moths Compiled by Everyone.
このサイトはみんなで日本産の蛾のWEB図鑑を作ろう！という趣旨で2003年11月5日に開設されました。

現在は更新されていません。

【お知らせ】
・2023年1月4日 掲示板を再設置しました。がいすとさん ありがとうございます。
・2022年12月23日 新しいハードウェア上にHTMLを復活させました。
・2022年12月17日 夕方にサーバーがすっ飛びました(壊れた)。いろいろ失われました。すみません。

【正誤表】
・2022年9月12日I巻の正誤追加。日本産蛾類標準図鑑I,II,IIIおよびIV巻の最新の正誤表はこちら発見時系列正誤表 巻別まとめI巻, II巻 , III巻, IV巻。
・2020年10月8日「日本の蛾」の最新の正誤表はこちら正誤表[Update!]

【掲示板】
【新・蛾像掲示板】 新しく設置しました

【縮小画像一覧】
・全成虫縮小画像一覧(注意：10,000枚以上あります！)
・全幼虫など縮小画像一覧

【日本産蛾類科一覧】
・和名をクリックすると種の一覧に、学名をクリックすると亜科や属の一覧になります。

| No. | 科和名 | 科学名 | 種数 | 成虫画像 | 幼虫など |
|---|---|---|---|---|---|
| 1 | コバネガ科 | Micropterigidae | 17 | 5 | 0 |
| 2 | スイコバネガ科 | Eriocraniidae | 5 | 1 | 0 |
| 3 | コウモリガ科 | Hepialidae | 9 | 7 | 2 |
| 4 | モグリチビガ科 | Nepticulidae | 65 | 2 | 0 |
| 5 | ヒラタモグリガ科 | Opostegidae | 8 | 0 | 0 |
| 6 | ツヤコガ科 | Heliozelidae | 21 | 2 | 0 |
| 7 | ヒゲナガガ科 | Adelidae | 35 | 20 | 1 |

**Fig. 1.7** The screenshot of web site of the identification guide of Japanese moths compiled by everyone (JPMOTH)

participated in Japanese local groups and do not want to display their photographs or observation locations in global community science programs. The number of species is relatively high compared to the number of observers and records in Japan, which indicates that data on Japanese endemic species are valuable for global community science. Since the Wild Bird Society of Japan (https://www.wbsj.org/en/), which is the Public Interest Incorporated Foundation of bird research and conservation in Japan, started operating eBird Japan in 2021 and planning events for eBird, the bird occurrence data in Japan will increase in eBird in the future.

Japanese ecologists are starting to contribute to global community science by protecting traditional community science programs (e.g., Osawa 2013) and promoting new ones in prefecture units or nationwide (e.g., Kadoya et al. 2009; Saito et al. 2015; Morii and Nakano 2017; Washitani et al. 2020; Suzuki-Ohno et al. 2017, 2023). Chapters 3, 4, 5, and 6 of this book introduce examples of ecological community science programs planned by ecologists in Japan. Chapters 7, 8, and 9 of this book introduce examples of ecological community science programs planned by NPO/NGO/PIIA and museums in Japan. In the following sections, I focus on the importance of science communication, data openness and closedness, and also decision-making and conservation activities for ecological community science programs.

## 1.4 Science Communication and Data Openness of Ecological Community Science

Science communication is the communication of scientific topics to nonexperts and is sometimes called outreach. Ecological community science programs can be considered a part of science communication (Chap. 10) and often require high science communication skills. If the aim of the program is environmental education and outreach, ecologists will be especially aware of science communication. However, if the aim is to collect data for scientific research, they may not be so. Even when the primary aim of the program is neither environmental education nor outreach, improving science communication skills will lead to collecting more accurate data and will facilitate collaboration with volunteers.

The content and methods of communication should be adaptable to participants from different educational levels and life circumstances. The content should be simple enough for nonexperts but accurate enough to avoid misunderstandings. The content of science communication is not limited to ecological information about target species but also includes other information, such as the procedure of scientific research, data utilization, data copyrights, and data management related to rare species. One-way communication is necessary to provide scientific information from experts, and two-way communication is necessary to increase the depth of understanding and the educational effects on participants (Chap. 10).

The openness of community science programs is closely linked to their contribution to scientific research. Ecologists strongly recommended it to the ecologists who planned community science programs. Openness mainly consists of (1) an open data management plan, (2) preregistration, (3) open data, (4) open codes, (5) open software, and (6) open-access publications. (1) An open data management plan is a formal document that describes how data will be collected, analyzed, stored, and shared during and after the project is completed. (2) Preregistration is the practice of producing a published document that describes the hypotheses to be tested, the methods, and the analysis to be undertaken. This makes the work more transparent and allows the ecologists to get feedback from their peers before research is undertaken. (3) Open data, (4) open codes, (5) open software, and (6) open-access publications are relatively simple requirements for ecologists because they are always recommended when a paper is published in an academic journal. Before starting an ecological community science program, ecologists should consider its openness.

The openness of community science programs is also closely linked to the satisfaction of participants. Reporting the organized data and the analytical results to the participants will greatly benefit those who want to increase their scientific knowledge or view data collected in the program. However, it is necessary to protect participants' data if their data are related to their personal information or if they just want to keep their data private. For example, participants may collect data in places related to their personal information such as their gardens and schools. Participants who are attached to observation sites may hesitate to publish the data of the sites unconditionally. At the same time, the copyrights of data from ecological community science programs, such as photographs, must be carefully protected and scrutinized from an ethical perspective. The program planner must respond to various individual cases. More information on openness is described in Chap. 11 of this book.

Openness in ecological community science programs is often gradual. The accuracy and type of the data released will depend on the viewer. The program planners and scientists will use raw data collected by participants for the analysis of the data. In some cases, participants may be able to watch raw data when collecting data in the programs. However, in other cases, low-resolution data may be released for the participants due to unavoidable reasons such as the protection of rare species. For outside parties (people who are not participants), only partial data or statistics may be released for some benefit of participants. The program planners need to set the accuracy and type of the data to be released for each type of viewer.

## 1.5 Decision-Making and Conservation Activities in Ecological Community Science

Community science includes decision-making regarding some local community issues based on scientific knowledge and analysis (Chap. 12). This may include conservation activities for eliminating alien species and preserving native species (e.g., Kondo et al. 2019). Decision-making and conservation activities with volunteers in ecological community science programs will become more important in the future because environments are changing dramatically and ecologists alone will not be able to investigate and protect biodiversity alone. It is important to know how to collaborate with volunteers in decision-making and conservation activities with volunteers.

Biodiversity and its conservation activities are generally affected by the features of national land. Biodiversity and its conservation activities in Japan are also affected by the features of the Japanese archipelago. One of the features of the Japanese archipelago is that it is surrounded by sea. The archipelago consists of four large islands (Hokkaido, Honshu, Shikoku, and Kyushu) and over 10,000 small islands including oceanic islands with unique ecosystems. Japanese ecologists must constantly check the invasion of alien species from overseas. Elimination of invasive alien species requires a great amount of time, effort, and budget. However, ecologists rarely have enough time, effort, and budget, and volunteers often work with them to eliminate invasive alien species (Chaps. 3, 9, and 13, Figs. 1.2 and 1.8).

**Fig. 1.8** An example of invasive alien species in Japan, *Bombus terrestris* (Chap. 13). Volunteers can distinguish *B. terrestris* from other bumble bee species because there are very few species that resemble *B. terrestris* (Photographer: M. Mitsuhata)

The monitoring and elimination of invasive alien species has become a major theme in ecological community science in Japan, and ecologists need to gain the understanding and cooperation of multiple stakeholders (Chap. 13).

Another feature of the Japanese archipelago is that the area of mountains occupies about 75% of the national land. The mountains are generally steep and mostly covered with forests. About two-thirds of the land area is forest, whereas one-third of the land area is used as residential areas. Since residential areas are small, the population is concentrated in big cities such as Tokyo. In the countryside, much nature remains, and residents and nature are very close and interact with each other. The mosaic landscape of natural, seminatural, agricultural, and residential lands in the countryside is often called "socio-ecological production landscape and seascape (SEPLS)" in English and "Satoyama and Satoumi" in Japanese (Fig. 1.9). SELPS is one of the important areas for maintaining biodiversity with spatial and temporal heterogeneity by intermediate disturbances such as traditional agriculture. Currently, local communities living in SELPS and maintaining the local ecosystem face aging and depopulation (Annual Report on Forest and Forestry in Japan (FY2020) 2020). Paddy and vegetable fields have been abandoned and are undergoing ecological succession. In addition to aging and depopulation, due to the depression of the forestry economy, old artificial forests have not been logged and have become dark forests.

**Fig. 1.9** An example of Socioecological Production Landscape and Seascape (SEPLS). The landscape where low mountains, small woodlands, residential areas, rice fields, rivers, and streams are adjacent are common in SEPLS. The right side is the rice field and the left side is the irrigation canal for supplying water to the rice field

There is a risk that the number of organisms that inhabit bright forests, grasslands, agricultural lands, and wetlands is decreasing in SEPLS (e.g., Katoh et al. 2009). As well as protecting the abundant nature, it is also an important mission for ecologists to maintain SELPS with volunteers (Chap. 14). More information on decision-making and conservation activities with volunteers are described in Chaps. 13 and 14 of this book.

## 1.6 Organization of this Book

I hope that readers understand the need to cover all of the above topics. This book consists of the following four parts.

Part I: What Is Community Science?
Part II: Ecological Community Science Programs
Part III: Considering Data and Environments with Participants
Part IV: Changing Environments with Participants

Part I provides basic information on community science (Chap. 2). Part II introduces four case studies of ecological community science programs planned by ecologists in Japan: Ant survey using specimens (Chap. 3), bee monitoring using photographs (Chap. 4), bird monitoring using videos and sounds (Chap. 5), and marine fish survey using environmental DNA (Chap. 6). These chapters will present examples of ecological community science programs using various methods and the advantages and disadvantages of each method. Part II also introduces two ecological community science programs planned by NPOs/NGOs/PIIAs and museums in Japan: Monitoring Site 1000 Project (Chap. 7) and Field Research of Dandelions in West Japan (Chap. 8). Chapter 9 reviews the relationship between ecological community science programs and museums in Japan. These chapters provide useful information about the experiences and connections of people participating in ecological community science activities. Part III consists of three chapters: Science communication (Chap. 10), databases and the rights of participants (Chap. 11), and decision-making in community science (Chap. 12). These chapters delve into the topics extracted from the findings of the case studies in Part II. The topics in these chapters are those ecologists may find less attractive, but they require careful consideration. Part IV introduces case studies involving alien species elimination (Chap. 13) and wetland restoration (Chap. 14) with participants. Conservation activities with participants are desirable as the final goal of ecological community science programs for conservation. Part IV provides valuable information about conservation activities. In the Editor's note, I have given my opinion as to the challenges of community science programs in Japan.

## References

Annual Report on Forest and Forestry in Japan (FY2020) (2020) Ministry of Agriculture, Forestry, and Fisheries. https://www.maff.go.jp/e/data/publish/attach/pdf/index-22.pdf

Biggs J et al (2015) Using eDNA to develop a national citizen science-based monitoring programme for the great crested newt (*Triturus cristatus*). Biol Conserv:18319–18328. https://doi.org/10.1016/j.biocon.2014.11.029

Delaney DG, Sperling CD, Adams CS, Leung B (2008) Marine invasive species: validation of citizen science and implications for national monitoring networks. Biol Invasions 10:117–128

Dickinson JL, Zuckerberg B, Bonter DN (2010) Citizen science as an ecological research tool: challenges and benefits. Annu Rev Ecol Evol Syst 41:149–172

Fontaine B et al (2012) New species in the old World: Europe as a frontier in biodiversity exploration a test bed for 21st century taxonomy. PLoS One 7(5):e36881. https://doi.org/10.1371/journal.pone.0036881

Henderson S, Ward DL, Meymaris KK. Alaback P, Havens K (2012) Project budburst: citizen science for all seasons. In: Dickinson JL, Bonney RE (eds) Citizen science: public participation in environmental research

Horn GV et al (2018) The iNaturalist species classification and detection dataset. In: Computer vision and pattern recognition, pp 8769–8778

Jinbo U, Suzuki T (2006) Possibility of taxonomy together with the internet: some on-line projects of moths. (in Japanese). TAXA 20:6–14

Kadoya T, Ishii HS, Kikuchi R, Suda S, Washitani I (2009) Using monitoring data gathered by volunteers to predict the potential distribution of the invasive alien bumblebee *Bombus terrestris*. Biol Conserv 142:1011–1017

Katoh K, Sakai S, Takahashi T (2009) Factors maintaining species diversity in *satoyama*, a traditional agricultural landscape of Japan. Biol Conserv 142:1930–1936

Kobori H et al (2016) Citizen science: a new approach to advance ecology education and conservation. Abstr Ecol Res 31(1):1-19. https://doi.org/10.1007/s11284-015-1314-y

Kondo Y et al (2019) Interlinking open science and community-based participatory research for socio-environmental issues. Curr Opin Environ Sustain 39:54–61

Kramer-Schadt S et al (2013) The importance of correcting for sampling bias in MaxEnt species distribution models. Abstr Divers Distrib 19(11):1366-1379. https://doi.org/10.1111/ddi.12096

Kullenberg C, Kasperowski D (2016) What is citizen science? – A Scientometric meta-analysis. PLoS One 11(1):e0147152. https://doi.org/10.1371/journal.pone.0147152

Larson ER et al (2017) Environmental DNA (eDNA) detects the invasive crayfishes *Orconectes rusticus* and *Pacifactacus leniusculus* in large lakes of North America. Hydrobiologia 800:173–185. https://doi.org/10.1007/s10750-017-3210-7

Losey J, Allee L, Smyth R (2012) The lost ladybug project: citizen spotting surpasses scientist's surveys. Am Entomol 58:22–24

Meyer RS et al (2021) The CALeDNA program: citizen scientists and researchers inventory California's biodiversity. Calif Agric 75:20–32

Miya M, Sado T, Oka S, Fukuchi T (2022) The use of citizen science in fish eDNA metabarcoding for evaluating regional biodiversity in a coastal marine region: a pilot study. Metabarcoding Metagenomics 6:133–144. https://doi.org/10.3897/mbmg.6.80444

Miyazaki Y et al (2014) Biological monitoring by citizens using web-based photographic databases of fishes. Biodivers Conserv 23:2383–2391

Miyazaki Y, Teramura A, Senou H (2016) Biodiversity data mining from Argus-eyed citizens: the first illegal introduction record of *Lepomis macrochirus macrochirus* Rafinesque, 1819 in Japan based on Twitter information. ZooKeys 569:123–133

Miyazaki Y, Murase A, Honda J, Yamaide J, Senou H (2020) Usefulness of a Japanese internet community for fish conservation. Biodivers Conserv 29:625–642

Morii Y, Nakano T (2017) Citizen science reveals the present range and a potential native predator of the invasive slug *Limax maximus* Linnæus, 1758 in Hokkaido, Japan. BioInvasions Records 6:181–186

Norouzzadeh MS et al (2018) Automatically identifying counting and describing wild animals in camera-trap images with deep learning. Signif Proc Natl Acad Sci 115(25). https://doi.org/10.1073/pnas.1719367115

Osawa T (2013) Monitoring records of plant species in the Hakone region of Fuji-Hakone-Izu National Park, Japan, 2001–2010. Ecol Res 28:541–541. https://doi.org/10.1007/s11284-013-1049-6

Phillips SJ et al (2009) Sample selection bias and presence-only distribution models: implications for background and pseudo-absence data. Ecol Appl 19(1):181–197. https://doi.org/10.1890/07-2153.1

Pocock MJO, Tweddle JC, Savage J, Robinson LD, Roy HE (2017) The diversity and evolution of ecological and environmental citizen science. PLoS One 12(4):e0172579. https://doi.org/10.1371/journal.pone.0172579

Saito K et al (2015) Utilizing the Cyberforest live sound system with social media to remotely conduct woodland bird censuses in Central Japan. Ambio 44:S572–S583

Sei Shonagon (n.d.) The Makura no Soshi (translated by McKinney, M (2007) The Pillow Book. Penguin Classics)

Silvertown J et al (2011) Citizen science reveals unexpected continental-scale evolutionary change in a model organism. PLoS One 6:e18927

Silvertown J et al (2015) Crowdsourcing the identification of organisms: a case-study of iSpot. Zookeys 480:125–146

Sullivan BL et al (2009) eBird: a citizen-based bird observation network in the biological sciences. Biol Conserv 142:2282–2292

Suzuki-Ohno Y, Yokoyama J, Nakashizuka T, Kawata M (2017) Utilization of photographs taken by citizens for estimating bumblebee distributions. Sci Rep 7:11215. https://doi.org/10.1038/s41598-017-10581-x

Suzuki-Ohno Y, Yokoyama J, Nakashizuka T, Kawata M (2022) Deep learning increases the availability of organism photographs taken by citizens in citizen science programs. Sci Rep 12:1210. https://doi.org/10.1038/s41598-022-05163-5

Suzuki-Ohno Y et al (2023) Evaluation of a community science monitoring with environmental DNA for marine fish species: "fish survey project using environmental DNA". Environ DNA 5:613–623. https://doi.org/10.1002/edn3.425

Swanson A et al (2015) Snapshot Serengeti, high-frequency annotated camera trap images of 40 mammalian species in an African savanna. Sci Data 2:150026. https://doi.org/10.1038/sdata.2015.26

Swanson A et al (2016) A generalized approach for producing, quantifying, and validating citizen science data from wildlife images. Conserv Biol 30:520. https://doi.org/10.1111/cobi.12695

Tøttrup AP et al (2021) Citizens in the lab: performance and validation of eDNA results. Citiz Sci 6:35

Van Eupen C et al (2022) Species profiles support recommendations for quality filtering of opportunistic citizen science data. Ecol Model 467:109910. https://doi.org/10.1016/j.ecolmodel.2022.109910

Washitani I, Nagai M, Yasukawa M, Kitsuregawa M (2020) Testing a butterfly commonness hypothesis with data assembled by a citizen science program "Tokyo Butterfly Monitoring". Ecol Res 35:1087–1094. https://doi.org/10.1111/1440-1703.12161

# Part I
# What Is Community Science?

# Chapter 2
# Fundamental Ideas and Concepts in Citizen Science

**Yuko Ikkatai**

**Abstract** Citizen science is a collaborative approach to scientific research that engages experts and the general public. It spans various fields, from natural sciences to the humanities. Projects vary according to objectives but continue to focus on research, education, or societal change, which frequently occur with trade-offs among these objectives. Several factors contribute to the advancement of citizen science, including the urgency of global environmental issues, expansion of big data, improved educational opportunities, policy recommendations, and technological advancements. Recently, artificial intelligence (AI) plays a vital role in enhancing the processing, scalability, and engagement of data, although ethical concerns must be addressed. Guiding principles and participation models provide essential frameworks for these initiatives. This chapter intends to provide basic knowledge of citizen science and its dynamic and diverse nature and global expansion.

**Keywords** Type · Goal · Citizen participation · Model · Japan

## 2.1 What Is Citizen Science?

Citizen science is currently spreading worldwide. Citizen science involves scientific research activities conducted through collaboration between experts, such as researchers, and the general public. For example, the Oxford English Dictionary (2014) defines citizen science as "scientific work undertaken by members of the general public, often in collaboration with or under the direction of professional scientists and scientific institutions." The term citizen in citizen science encompasses all individuals who may be experts in one field but not in others. Therefore, in the context of citizen science, citizens potentially include everyone. Citizen

Y. Ikkatai (✉)
Institute of Human and Social Sciences, Kanazawa University, Kanazawa-shi, Ishikawa, Japan
e-mail: y.ikkatai@staff.kanazawa-u.ac.jp


Y. Suzuki-Ohno (ed.), *Community Science in Ecology*, Ecological Research Monographs, https://doi.org/10.1007/978-981-97-0304-3_2

science covers a wide range of fields, including not only the natural sciences but also the humanities, social sciences, and arts. SciStarter (https://scistarter.org/), which is an online platform for citizen science, offers a search feature that enables users to find projects under various categories. It features 1407 projects in 28 topics available as of August 31, 2023. These projects span from agriculture and animals to the social sciences and science policy, which showcases the diversity of subjects covered. Another online platform, Zooniverse (https://www.zooniverse.org/), lists 97 projects across 11 fields, including arts, biology, climate, history, language, literature, medicine, nature, physics, social science, and space, as of August 31, 2023.

Citizen science is a concept that was nearly simultaneously proposed in the 1990s by Rick Bonney in the United States and Aran Irwin in the United Kingdom. Sociologist Aran Irwin defined citizen science as a process in which science should respond to the interests of citizens, whereas citizens should advance and enhance the creation of reliable scientific knowledge. Alternatively, Rick Bonney, an ornithologist, presented citizen science as a process in which nonscientist citizens, such as bird watchers, collect data based on scientific procedures (e.g., Lewenstein 2022). Cooper and Lewenstein (2016) refer to the former as democratic and the latter as contributory, but they note that these concepts are relatively complementary aspects of citizen science instead of mutually exclusive (Haklay et al. 2023).

Dozens of similar terms for citizen science have also, such as amateur science, crowdsourced science, volunteer monitoring, and public participation in scientific research. Recently, the term community science has frequently been used in place of citizen science, mainly in the United States (Cooper et al. 2021). Citizen and community science involve public participation in science, but citizen science can be perceived as excluding people who do not hold citizenship in a specific country or conveying hostility toward them (Lowry and Stepenuck 2021). Community science places a stronger emphasis on diversity, equity, and inclusion and encourages a wide range of demographic. The Citizen Science Association (CSA) in the United States announced the change of its name to an Association Advancing Participatory Sciences in May 2023. CSA expresses that the final "s" in sciences is important, because it emphasizes the small one that affirms a plurality of people and practices (CSA 2023). Many terms refer to citizen science, but Eitzel et al. (2017) argue that they generally share characteristics similar with those of the concepts of democratic and/or contributory citizen science.

Citizen science denotes the harnessing of the collective effort of many, which leads to discoveries that a lone scientist may not achieve. SciStarter highlights four common characteristics shared by citizen science projects, namely, (1) anyone can participate; (2) participants use the same protocol, such so data can be combined and be of high quality; (3) data can help real scientists draw realistic conclusions; and (4) a wide community of scientists and volunteers work together and share data to which the public, as well as scientists, have access (SciStarter n.d.). An example of such projects is eBird (https://ebird.org/home) in which individuals capture photographs of local birds and collaboratively create a large-scale database. The collected data are made accessible to citizens and researchers and are utilized for conservation and ecological understanding. Additional examples include Galaxy

Zoo (https://www.zooniverse.org/projects/zookeeper/galaxy-zoo/), in which participants collectively classify the shapes of numerous galaxies from images on the Internet, and Shakespeare's World (https://www.shakespearesglobe.com/discover/shakespeares-world/), in which volunteers transcribe historical documents from Shakespeare's era into digital data.

The goals of citizen science projects can significantly vary across projects. Kobori (2023) broadly categorizes the goals of citizen science into three areas, namely, research, education, and societal change. When research is the primary goal, the focus lies on encouraging citizen participation in scientific research and promoting open science practices. In such cases, citizen science projects intend to engage the public in scientific research activities and facilitate the advancement of open science principles. If education is the primary objective, then emphasis shifts toward increasing interest in nature or society and serving as a lifelong learning opportunity. These projects aim to increase the awareness of participants of scientific concepts and provide hands-on learning experiences. In cases in which societal change is the goal, the focus is on addressing real-world challenges such as problem-solving and community revitalization. These projects endeavor to contribute to positive changes in society and actively involve communities in addressing local issues. However, recognizing that these goals frequently involve trade-offs and simultaneously achieving all objectives can be challenging for a single project are essential aspects. Therefore, considering in advance which of the three goals should be prioritized and highlighted while remaining aware of the potential trade-offs among them is crucial for project organizers (Kobori 2023).

The number of citizen science projects that are primarily focused on environmental and ecological topics has been increasing steadily over the years (Pocock et al. 2017). Additionally, the number of peer-reviewed papers that summarize the results of these projects is also on the rise (Follett and Strezov 2015). Citizen science is gradually becoming increasingly extensive; at the same time, however, the formats of participation vary widely, which range from complex to straightforward (Pocock et al. 2017). Alternatively, citizen science is not necessarily a one-size-fits-all solution for all research fields. For example, it is highly suitable for research areas that involve intensive data collection or analysis of large datasets. However, it may not be well suited for fields that require advanced specialized knowledge or that rely on large-scale equipment and specialized laboratories (Pocock et al. 2014).

Kobori (2023) points out several factors that contribute to the increasing recognition of citizen science. The author states that these factors highlight the increased reliance on citizen involvement for addressing environmental and societal challenges and outlines six key points. First, the severity of global environmental issues has been escalating, which demands societal responses. These issues cannot be solved solely by scientists or government officials, such that citizen engagement has become essential. Second, the proliferation of big data has amplified the significance of citizen science. The ability to collect vast amounts of data, which was previously unattainable and transcended national boundaries, underscores the importance of citizen participation. Third, an improvement has been noted in educational opportunities. Participation in citizen science enables individuals to contribute to scientific endeavors and enhance their knowledge and understanding of science. Fourth, policy recommendations have been emerging as an outcome of efforts toward citizen science. Recognizing that policy recommendations that stem from the findings of citizen science typically gain social consensus and are more easily incorporated into policy decisions. Fifth, the widespread availability of tools, such as the Internet, mobile devices, and detection equipment, has opened doors for citizen involvement in research processes, that were traditionally exclusive to experts. Sixth, changes in the social environment have exerted their impact. Notably, the advancement of levels of education and the concept of open science have gained prominence. Open science involves making of research data and outcomes publicly available for broader societal use, because they were previously accessible only to researchers. Thus, citizen science can be viewed as an endeavor to share the research process with the public.

Another recent contribution to citizen science is the use of AI, which has rapidly expanded. AI is not only used for verifying the quality of data collection and data processed by citizens but also for providing information tailored to the interests of participants to sustain their engagement. It is also employed for real-time feedback, among others (Lotfian et al. 2021). Ceccaroni et al. (2023) outline several benefits of incorporating AI in citizen science, including (1) increasing the speed and scale of data processing, (2) increasing the temporal and geographical scopes of projects, (3) improving the quality of data collected and processed, (4) supporting learning between humans and machines, (5) leveraging new data sources, and (6) diversifying engagement opportunities. Of particular note is the rapid processing of large-scale data, which significantly contributes to scientific advancement. AI enables the processing of vast amounts of data in a more accurate and swift manner than manual methods can. AI proves to be an extremely valuable tool for projects aimed at achieving scientific outcomes. In the future, effectively harnessing AI in citizen science holds the potential to diversify opportunities for scientific engagement, promote equal educational opportunities, and bridge the gap between individuals with varying levels of interest in science and between experts and nonexperts. Expectations for the use of AI in citizen science are high; however, concerns are gradually emerging regarding its implementation. These concerns cover a wide range of issues, including data breach, inappropriate commercial use of data, accountability of project organizers, method of obtaining consent, and data bias related to gender and race, among others (Ceccaroni et al. 2019).

## 2.2 Guideline and Models of Citizen Participation

The perspective of citizen participation is crucial in citizen science. The European Citizen Science Association (ECSA) has proposed the "10 Principles of Citizen Science" as essential guidelines for excellent citizen science (Table 2.1, ECSA 2015). These principles have been translated into 38 languages, including Japanese (as of August 2023), and widely shared worldwide. They emphasize aspects of

**Table 2.1** Ten principles of citizen science

| | |
|---|---|
| 1 | Citizen science projects actively involve citizens in scientific endeavor that generates new knowledge or understanding. Citizens may act as contributors, collaborators, or as project leader and have a meaningful role in the project |
| 2 | Citizen science projects have a genuine science outcome |
| 3 | Both the professional scientists and the citizen scientists benefit from taking part |
| 4 | Citizen scientists may, if they wish, participate in multiple stages of the scientific process |
| 5 | Citizen scientists receive feedback from the project |
| 6 | Citizen science is considered a research approach like any other, with limitations and biases that should be considered and controlled for |
| 7 | Citizen science project data and meta-data are made publicly available and where possible, results are published in an open access format |
| 8 | Citizen scientists are acknowledged in project results and publications |
| 9 | Citizen science programs are evaluated for their scientific output, data quality, participant experience, and wider societal or policy impact |
| 10 | The leaders of citizen science projects take into consideration legal and ethical issues surrounding copyright, intellectual property, data sharing agreements, confidentiality, attribution, and the environmental impact of any activities |

citizen participation such as actively involving citizens (Principle 1), allowing citizens to participate in various scientific processes (Principle 4), and ensuring that citizens receive feedback on outcomes (Principle 5).

Scholars have proposed several models for citizen participation. An example is the concept of the "Ladder of Citizen Participation" by Sherry Arnstein, which categorizes the levels of citizen engagement into eight rungs: "manipulation," "therapy," "informing," "consultation," "placation," "partnership," "delegated power," and "citizen control" (Arnstein 1969). The bottom rung represents formal participation, which does not necessarily involve substantial citizen engagement. As citizens ascend the ladder, the level of participation increases, and they become more empowered. At the top rung (citizen control), citizens take an active role in managing citizen participation.

Shirk et al. (2012) divided citizen participation in scientific research into five categories based on the level of citizen involvement, namely, contractual, contributory, collaborative, co-created, and collegial. Contractual projects involve communities that hire professional researchers to perform a particular scientific investigation and provide a report on the findings. Scientists typically create contributory projects with the primary role of the public lying in the contribution of data. Scientists frequently initiate collaborative projects in which members of the public not only contribute data but also assist in refining project design, analyzing data, and/or sharing findings. Co-created projects are developed through collaboration between scientists and the public with active participation from at least a few members of the public in the majority, if not all, aspects of the research process. In collegial projects, individuals without formal credentials conduct independent research in which the level of expected recognition by institutionalized science and/or professionals varies.

Haklay (2013) categorized the levels of participation in citizen science into four levels. In Level 1 (crowdsourcing), citizens serve as sensors for data collection and primarily contribute to this process. In Level 2 (distributed intelligence), citizens take on roles in interpreting and making judgments about the data. In Level 3 (participatory science), citizens are engaged in problem formulation, but experts conduct data analysis. In Level 4 (extreme citizen science), citizens take the lead and participate in problem formulation and data collection and analysis. Notably, however, a higher level is not necessarily better. Depending on the objectives of the project, selecting an appropriate level or a combination of levels may be more suitable.

Ideally, participating in citizen science is desirable for more people, but it should not denote mandatory involvement. Depending on the project, individuals may be well suited for it, and others may not. Skarlatidou and Haklay (2021) used DITO's escalator model to explain participation in science, including citizen science, in seven stages. This model categorizes various activities based on the scale of participant involvement. For example, the ground floor, Level 1 (everyone), involves activities in which the entire population participates. Level 2 (passive consumption of science) represents the passive consumption of science such as watching television programs. Level 3 (active consumption of science) involves the active consumption of science such as visiting museums. Level 4 (opportunistic or highly limited participation) entails limited but active engagement such as birdwatching. Level 5 (joining volunteer computing or thinking) involves participation in volunteer computing or thinking, such as contributing to online projects, such as Zooniverse. Level 6 (data collection and analysis) denotes continuous data collection or processing, including providing data to platforms such as eBird. Finally, Level 7 (high engagement in DIY science) is do-it-yourself science in which individuals develop their sensors and collect data. The number of participants tends to decrease with the increase in levels. The authors defines citizens involved in Levels 5 through 7 as serious citizen scientists and estimates that more than 1.7 million of them exist. Notably, high levels are not inherently better, and individuals can move up or down the scale according to their interests and life stages. For instance, they may temporarily descend due to childcare or caregiving responsibilities or ascend after retirement, when they have more time. Creating an environment that enables such diverse choices is desirable in citizen science. Skarlatidou and Haklay (2021) emphasize the need for appropriate funding, support, and attention for the growth of citizen science.

## 2.3 Who Is It Suitable for Citizen Science?

The suitability for citizen science varies on the basis of individual interests, skills, and circumstances. For instance, those with an interest in science or the environment may find citizen science appealing. Citizen science frequently involves scientific research activities, and many projects revolve around environmental issues. It is also well suited for individuals who excel in cooperation and communication. Projects, such as Galaxy Zoo, even when conducted online, typically facilitate interaction among participants. Thus, individuals who wish to contribute to their local communities may also find it a good fit. Citizen science frequently involves community-based activities that enable individuals to contribute to their local surroundings. Additionally, they should consider whether or not they can commit the necessary time and resources, especially to larger projects that require regular participation. Alternatively, engaging individuals with less interest in science can be challenging. Limited time or resource can also pose obstacles to participation. A few projects may demand advanced expertise, which renders them unsuitable for general participants. Citizen science encompasses a wide array of projects, such that identifying projects aligned with one's interests and skills is crucial. Ultimately, finding projects that resonate with one's passion and ability is essential.

## 2.4 Citizen Science in Japan

Citizen science is thriving in, for example, Europe, the United States, and Australia with support from policy initiatives. In addition, in Japan, the policy momentum to support citizen science is increasing. The Fifth Science and Technology Basic Plan, which covers the period from the fiscal year 2016 to 2020 (Ministry of Education, Culture, Sports, Science, and Technology 2016), includes a provision that states, "Efforts to promote citizen science and create methodologies and environments for researchers to plan research projects, conduct research, and disseminate results in collaboration with the general public and policymakers should be promoted." Similarly, the Sixth Science and Technology Innovation Basic Plan, which spans from fiscal year 2021 to 2025 (Ministry of Education, Culture, Sports, Science and Technology 2021), emphasizes the resolution of social issues through citizen participation and the activation of citizen science. It specifically mentions initiatives as follows:

> … launching citizen science research projects that involve a large number of citizens (anticipated to reach tens of thousands by fiscal year 2022) for tasks like collecting numerous samples and conducting scientific experiments. This reflects a bottom-up approach involving diverse stakeholders in academia, industry, and government in the process of shaping new science, technology, and innovation policies.

Citizen science in English can be expressed using katakana as "シチズンサイエンス" (shitizunsaiensu) and kanji as "市民科学" (shiminkagaku) in Japanese. Although a strict demarcation does not exist between these terms, they carry different historical connotations. Citizen involvement in science and technology in Japan can be traced back to the 1960s, when active engagement in addressing issues, such as pollution, health hazards, and environmental destruction, was observed. This was characterized by grassroots investigation conducted by citizens for the sake of citizens, which was a concept akin to "citizens conducting science for citizens." Subsequently, from the 1970s onward, citizen-led investigation and environmental NGOs emerged along with antinuclear movements (van Oudheusden et al. 2020;

Kobori 2023). The concept of shiminkagaku was introduced only in 1999 by Jinzaburo Takagi, the founder of the citizens' organization called Citizens' Nuclear Information Center. He popularized the idea of "science by the hands of citizens." In this context, citizen science became associated with the criticism of institutionalized science and science policies (van Oudheusden et al. 2020). Therefore, shiminkagaku carries the implication of a movement for democratizing science in which citizens gain sufficient strength to address significant societal issues as a counterbalance to the existing power structure. Alternatively, shitizunsaiensu is a relatively newer term, which primarily emphasizes citizen participation in scientific research. It frequently pertains to scenarios in which citizens collaborate with professional scientists, particularly for large-scale data collection and analysis (Yoshizawa 2018). Many contemporary researcher-led projects in Japan fall more under the category of shitizunsaiensu. Additionally, in Japanese government documents, such as basic plans for science and technology, the term shitizunsaiensu is used to refer to citizen involvement in scientific activities (Ministry of Education, Culture, Sports, Science and Technology 2016, 2021).

In recent years in Japan, nascent citizen science projects have begun to yield results. Individuals researchers or research groups operate a few of them. For instance, the *Finding Slugs Project* https://sites.google.com/site/udakawebsite/madarakouranamekuji-limax-maximus) conducts a nationwide survey of invasive slugs in Japan. The *Hanamaru-Maruhana National Census* (http://hanamaruproject.s1009.xrea.com/hanamaru_project/index.html) surveys the distribution of bumble bees across Japan. Another example is the *#KantoSnowCrystal project* (https://www.mri-jma.go.jp/Dep/typ/araki/snowcrystals.html), which collects photographs of snowflakes. These projects rely on data collection through personal devices such as smartphones. Other projects such as the *Thundercloud Project* for collecting lightning data or *Safecast* (https://safecast.jp/) for environmental radiation data involve the use of specific measurement devices provided by project organizers. Additionally, other projects, such as *Minna De Honkoku* (https://honkoku.org/), transcribe historical texts through the help of volunteers using a data-processing approach. Furthermore, academic societies and research institutions operate certain projects. For instance, the National Astronomical Observatory of Japan hosts *Galaxy Cruise* (https://galaxycruise.mtk.nao.ac.jp/), which focuses on classifying galaxy images. Interestingly, Galaxy Cruise opts to avoid the term citizen science and uses the term citizen astronomy. The Japanese Psychological Association collaborated with certified psychologists to conduct a *Citizen Science Project* (https://psych.or.jp/authorization/citizen/). These diverse initiatives showcase the various forms and approaches that citizen science projects take in Japan.

Furthermore, nonprofit organizations (NPOs) and businesses operate a number of "citizen science" projects in Japan. For instance, the NPO Bird Research (http://www.bird-research.jp/) manages several citizen-participation bird surveys. In 2019, the company BIOME released the *Biome* app (https://biome.co.jp/app-biome/), which is equipped with an AI that identifies species names with high accuracy when users take photos of flora and fauna. Users can select the most likely species name and contribute data to the database. Weather News Inc. (https://jp.weathernews.

com/), which is a meteorological company, encourages app users to take photos of the sky, which are then used to improve the accuracy of weather forecasting. Notably, when NPOs or businesses operate these projects, they may not always explicitly use the term citizen science. Thus, debate arises about whether or not activities conducted by businesses should be defined as citizen science; nevertheless, their activity remains very similar to citizen science. In certain cases, researchers collaborate with NPOs or companies that own data collection systems and use them to collect valuable research data.

## 2.5 Conclusions

In this chapter, the author reviewed the definition, concept, levels of citizen science, and recent notable activities associated with it. Additionally, the emergence of promising projects in Japan that have begun to yield results was introduced. However, in a survey conducted in 2018 that targeted 1034 men and women living in Japan, only 1% of the respondents reported a prior experience of participating in citizen science, whereas 19% expressed an interest in participating (Ikkatai et al. 2020). At present, it's challenging to claim that citizen science is widely pervasive in Japan. However, there is a potential for government support in the future, which may lead to an increase in the number of citizen science projects. It is crucial to clarify the purpose of conducting citizen science while also ensuring ethical considerations are not overlooked.

**Acknowledgments** Author thanks to the Drs, Y. Suzuki-Ohno and T. Osawa for discussion about the practice of citizen science. This work was supported by JSPS KAKENHI Grant Number 20K14114.

## References

Arnstein SR (1969) A ladder of citizen participation. J Am Inst Plann 35(4):216–224. https://doi.org/10.1080/01944366908977225

Ceccaroni L, Bibby J, Roger E, Flemons P, Michael K, Fagan L, Oliver JL (2019) Opportunities and risks for citizen science in the age of artificial intelligence. Citiz Sci 4(1):29. https://doi.org/10.5334/cstp.241

Ceccaroni L, Oliver JL, Roger E, Bibby J, Flemons P, Michael K, Joly A (2023) Advancing the productivity of science with citizen science and artificial intelligence. In: Artificial intelligence in science: challenges, opportunities and the future of research. OECD Publishing

Cooper CB, Lewenstein BV (2016) Two meanings of citizen science. In: Kennedy EB, Cavalier D (eds) The rightful place of science: citizen science. Consortium for science, policy, & outcomes, pp 51–62

Cooper CB, Hawn CL, Larson LR, Parrish JK, Bowser G, Cavalier D, Dunn RR, Haklay MM, Gupta KK, Jelks NO, Johnson VA, Katti M, Leggett Z, Wilson OR, Wilson S (2021) Inclusion in citizen science: the conundrum of rebranding. Science 372(6549):1386–1388. https://doi.org/10.1126/science.abi648

CSA (2023) Announcing a new name for this association. https://citizenscience.org/2023/07/14/announcing-a-new-name-for-this-association/. Accessed 31 Aug 2023

ECSA (2015) Ten principles of citizen science. https://www.ecsa.ngo/documents/. Accessed 17 Aug 2023

Eitzel MV, Cappadonna JL, Santos-Lang C, Duerr RE, Virapongse A, West SE, Kyba CCM, Bower A, Cooper CB, Sforzi A, Metcalfe AN, Thiel M, Haklay M, Ponciano L, Roche J, Ceccaroni L, Shilling FM, Dörler D, Heigl F, Kiessling T, Davis BY, Jiang Q (2017) Citizen science terminology matters: exploring key terms. Citiz Sci 2:1–20. https://doi.org/10.5334/cstp.96

Follett R, Strezov V (2015) An analysis of citizen science based research: usage and publication patterns. PLoS One 10:e0143687. https://doi.org/10.1371/journal.pone.0143687

Haklay M (2013) Citizen science and volunteered geographic information: overview and typology of participation. In: Sui D, Elwood S, Goodchild M (eds) Crowdsourcing geographic knowledge. Springer, pp 105–122

Haklay M, König A, Moustard F, Aspee N (2023) Citizen science and post-normal science's extended peer community: identifying overlaps by mapping typologies. Futures 150:103178. https://doi.org/10.1016/j.futures.2023.103178

Ikkatai Y, Ono E, Udaka H, Enoto T (2020) Relationship between the level of interest in science and technology and the willingness to participate in citizen science in Japan (in Japanese). Japanese J Sci Commun 27:57–70. https://doi.org/10.14943/95029

Kobori H (2023) Citizen science recommendations (in Japanese). Bun-ichi Sogo Shuppan, Tokyo

Lewenstein BV (2022) Is citizen science a remedy for inequality? Ann Am Acad Pol Soc Sci 700:183–194. https://doi.org/10.1177/00027162221092б

Lotfian M, Ingensand J, Brovelli MA (2021) The partnership of citizen science and machine learning: benefits, risks, and future challenges for engagement, data collection, and data quality. Sustainability 13:8087. https://doi.org/10.3390/su13148087

Lowry CS, Stepenuck KF (2021) Is citizen science dead? Environ Sci Technol 55:4194–4196. https://doi.org/10.1021/acs.est.0c07873

Ministry of Education, Culture, Sports, Science and Technology (2016) The 5th science and technology basic plan (in Japanese). https://www8.cao.go.jp/cstp/kihonkeikaku/5honbun.pdf. Accessed 17 Aug 2023

Ministry of Education, Culture, Sports, Science and Technology (2021) The 6th science, technology and innovation basic plan (in Japanese). https://www8.cao.go.jp/cstp/kihonkeikaku/6honbun.pdf. . Accessed 17 Aug 2023

Oxford English Dictionary (2014) Citizen science. Oxford University Press, Oxford

Pocock MJ, Chapman DS, Sheppard LJ, Roy HE (2014) Choosing and using citizen science: a guide to when and how to use citizen science to monitor biodiversity and the environment. NERC/Centre for Ecology & Hydrology

Pocock MJ, Tweddle JC, Savage J, Robinson LD, Roy HE (2017) The diversity and evolution of ecological and environmental citizen science. PLoS One 12:e0172579. https://doi.org/10.1371/journal.pone.0172579

SciStarter (n.d.) What is citizen science? https://scistarter.org/citizen-science. Accessed 31 Aug 2023

Shirk JL, Ballard HL, Wilderman CC, Phillips T, Wiggins A, Jordan R, McCallie E, Minarchek M, Lewenstein BV, Krasny ME, Bonney R (2012) Public participation in scientific research: a framework for deliberate design. Ecol Soc 17:29

Skarlatidou A, Haklay M (2021) Citizen science impact pathways for a positive contribution to public participation in science. J Sci Commun 20:A02. https://doi.org/10.22323/2.20060202

Van Oudheusden M, Kenens J, Yoshizawa G, Mizushima N, Van Hoyweghen I (2020) Developing pathways for nuclear science, technology and innovation governance (in Japanese). J Sci Technol Stud 18:58–73. https://doi.org/10.24646/jnlsts.18.0_58

Yoshizawa G (2018) What is citizen science? (in Japanese). https://www.shiminkagaku.org/csijnewsletter_045_201805_yoshizawa_02/. Accessed 17 Aug 2023

# Part II
# Ecological Community Science Programs

# Chapter 3
# Practicing Collaborative Biodiversity Research with High Schools to Benefit Both Education and Invasive Species Monitoring

**Masashi Yoshimura**

**Abstract** We discuss here the challenges of designing research projects involving people from different backgrounds whose primary purpose is not necessarily to generate academic outputs, based on our experience in coordinating a community-based biodiversity monitoring project for ants. Local ant faunal research is active in Japan, and there is an accumulation of robust identification tools, but several barriers exist to developing sustainable models for citizen participation. For example, there are gaps between the planners, often researchers at the top of the power hierarchy, and participants, citizen scientists with varying levels of experience, regarding their motivation, willingness to participate, and sense of purpose. In this example involving collaboration with local high schools, there were large discrepancies in survey cost, data accuracy, and expected results between researchers who mainly wanted to discover new scientific findings and schools who mainly prioritized educational value of activities.

To address these issues, we attempted a new form of networking by having each participating organization independently design and implement their project. We did this by sharing ant survey packages, which enabled unified and compact analyses implementing the Time-Unit Sampling method. By sharing the same survey protocols, we ensured high data quality by sharing the specimens themselves, rather than the data. This framework helped us demonstrate a good example of win-win project collaboration that allowed the participating schools to conduct their projects autonomously while specimens collected by students could be utilized for scientific research. Thus, this framework benefited both high school education and local biodiversity monitoring research that informs the distribution status and control of invasive ant species.

**Keywords** Cooperation · Collection · Conservation · Formicidae · Environmental education · Fire ant · Okinawa · Taxonomy

M. Yoshimura (✉)
Okinawa Institute of Science and Technology Graduate University, Okinawa, Japan
e-mail: myoshimura@ant-database.org



Y. Suzuki-Ohno (ed.), *Community Science in Ecology*, Ecological Research Monographs, https://doi.org/10.1007/978-981-97-0304-3_3

## 3.1 Introduction

### 3.1.1 Objective Setting and Collaborative Design for Working with Citizens in Natural History Research

Many of the readers of this book will be expected to be involved in citizen science from the research side. The research side here refers to the role of planning and managing projects that include citizen science elements, collecting and analyzing data collected with citizen help, and presenting the final results. In this chapter, I would like to discuss the design and challenges of research projects involving people from diverse backgrounds whose primary purpose is not necessarily to produce academic output, based on our experience in coordinating a regional biodiversity monitoring project. When planning a natural history or ecological research project on living organisms, it is first necessary to select appropriate targets and data collection methods in line with the objectives, temporal and spatial scales, and analytical methods of the data. Those appropriate choices regarding these will be commonly required to ensure the quality of research results, regardless of whether the research project includes a citizen science component or not. First, with regard to the objective setting, the design will have significant differences depending on whether the project seeks the participation of the public in the collection of large-scale data for pure research purposes, or whether it is intended to be an environmental educational opportunity aiming toward long-term, active citizen involvement in local environmental conservation. The temporal scale chosen will make a difference in the data collection, governing whether the data collection should be conducted as a one-time event capturing a condition at a single point in time, or whether it should be designed to reveal dynamics across seasons or years. An example of the former would be a study on how many flowers of a certain species are blooming in a forest, while an example of the latter would be to study whether the timing in blooming of these plants is affected by global warming. The spatial scale chosen can be the difference between focusing on the entire globe, a region, or a single forest. Ideally, the study organisms should be selected based on these temporal and spatial scales, but there are many other considerations, at least in the case of citizen science, where nonspecialists are expected to participate in the data collection process. For example, what levels of burden and expertise are required to collect data of sufficient quantity and quality, and how should the necessary cooperation of citizens be obtained? In terms of citizen participation, it would be ideal if the survey targets are also familiar to nonspecialists, are easily identifiable, lead to solutions to local issues, and meet the needs of the citizens.

### 3.1.2 Advantages and Challenges of Insects as Research Materials

Insects are often chosen as familiar research targets in citizen-engaged research design, along with birds and flowers that are popular with many citizen enthusiasts. In comparison to birds, insects are generally characterized by small body size, small home ranges, large numbers of individuals, high population densities, high diversity, and high environmental dependence on microhabitats. Despite public familiarity with insects, some of the above characteristics are often considered difficulties in using insects as research materials. First, their small body size may sometimes require special tools such as loupes or stereomicroscopes and forceps for observation. Next, knowledge and skills for sample identification often limit research progress due to the high diversity of insects. Furthermore, high population densities and large sample sizes of collections can require much time for sample processing and space for storage. But on the other hand, I believe that the attractiveness of insects as material for collaborative research with the public is more than offset by these potential difficulties. Their small body size and small home ranges allow the construction of research designs with a smaller restricted area, even in an isolated forest remaining in urban areas. High population densities leading to large sample sizes allow for greater statistical power and flexibility in the analysis of collected data. High diversity and microhabitat dependence mean high potential for use as environmental indicator organisms, making them suitable material for setting up themes for citizen collaboration such as environmental assessment of familiar areas. In addition to these advantages, using insects can provide a solution for one of the challenges in a research design, which relies on citizen participation rather than the researchers themselves to collect data, to ensure the accuracy of the collected data, because insects can be collected and stored as specimens easily. Thus, insects also can be a suitable target from also the standpoint of verifiability of data and analysis results.

### 3.1.3 Advantages and Challenges of Ants as a Research Target

I discussed above the advantages and difficulties of citizen science using insects as research materials. For citizen science, it seems that butterflies and bees are often used because of the ease of finding, observing, and collecting specimens, as well as for identification. In addition, rather than dealing with communities that require the identification of many species, the distribution, movement, and population dynamics of specific species are often chosen as themes. Because ants are small compared to butterflies and bees, the need for collecting specimens for species identification and using stereomicroscopes are issues that need to be resolved when working with the public. However, ants have many advantages over other organisms if these issues can be resolved to some extent in the research design. The ant is one of the most

**Fig. 3.1** Ants designated as invasive alien species in Japan. Left, the red imported fire ant *Solenopsis invicta* Buren, 1972 from Taiwan; right, the browsing ant *Lepisiota frauenfeldi* (Mayr 1855) from Okinawa-Jima

familiar insects to people, occupying a dominant position in all ecosystems except in some cold regions. From forested to urban environments, ants are probably one of the first creatures you see when you look down at your feet. As they live in large social groups and have high population densities, ants are an essential component in a variety of niches in the ecosystem, whether as predators or prey species for other organisms, plant dispersers, decomposers, or scavengers. The features of worker ants, which are wingless and forage on the ground or on plants, also facilitate observation and collection for nonexperts. Their social behavior in constructing colonies makes it safe to assume that the collection of an individual exerts virtually no collection impact on that population. These features of ants provide the potential to treat whole fauna as themes rather than just distributions or populations focused on a specific species. In addition to these advantages, many invasive alien species are well known by the public, such as fire ants *Solenopsis* spp. and the Argentine ant *Linepithema humile* (Mayr 1868), which have become increasing threats due to expanding global trade and can significantly disturb the ecosystems they invade (Fig. 3.1). It is fair to say that ant research is in need of a citizen-participatory research model, not only because of public familiarity and the potential of more flexible target setting but also from the aspect of the social issue of monitoring invasive alien species and their control.

## 3.2 Utilization of Ants in Environmental Education

In Japan, ants are used often in educational settings, and one of the main reasons for their popularity is the progress in taxonomic information on this group in Japan. Of the approximately 300 species of ants recorded so far in Japan, more than 90% have been identified or described, and Japanese names have given for distinguishing even those with undetermined scientific names. Due to the enormous diversity of insects, there are still many groups for which such taxonomic knowledge has not yet been developed. Regardless of whether the research focuses on a single population of a particular species or a community of several species, taxonomic information about the group is an essential and basic tool. No matter how attractive a taxon may be, without basic information on the taxon, it is difficult to use as a research target. Such basic information includes how many species are in the area, how many are related or similar to the target species, and how they can be correctly distinguished from other similar species. In this sense, ants are a relatively easy insect taxon to work with in Japan because of the accessibility of such taxonomic information. Of course, the information infrastructure for utilizing ants for research was not developed overnight. The following is a brief history of the development of this information infrastructure in Japan.

### 3.2.1 The History of the Faunal Research on Japanese Ants and the Contribution of the Japanese Ant Image Database

The current wellspring of knowledge on Japanese ant fauna is undoubtedly the fruit of the tireless efforts of the Myrmecological Society of Japan and others in various capacities. The rapid progress of this organization can be attributed to a series of publications from 1988 to 1994 (Myrmecological Society of Japan 1988, 1989, 1991, 1992; Terayama and Kihara 1994) that consolidated the previously scattered knowledge on the taxonomy and distribution of Japanese ants. Through this series of works, all known species were reconsidered and given Japanese names, and unidentified species were given united species codes. These rapidly led to new discoveries and descriptions and developed the basis for subsequent ant research in Japan. The release of the Japanese Ant Image Database (Japanese Ant Database Group 1995) the following year was also a major event in promoting the use of ants in education and research (Fig. 3.2). Opening this website was a milestone, as it digitized the series of publications of Japanese ant fauna and taxonomy until 1994 and included color images of specimens for all species. This website was also designed with a strong emphasis on educational use, and an attempt was made to provide a pictorial identification page. With the expanding distribution of Internet technology, the Japanese ant image database grew and was updated repeatedly (Japanese Ant Database Group 2008 2008; Japanese Ant Database Group 1998, 2003a), providing an entry point for many people to study ants for various purposes, from research to education. Furthermore, since the database group published illustrated books of Japanese ants in 2003 (Japanese Ant Database Group 2003b, c), various books on ant identification have been published in Japan, ranging from regional to countrywide, and specialized to general field guides (e.g. Terayama et al. 2009, 2014; Yamane et al. 1999, 2010; Yoshizawa and Mizoguchi 2019). These

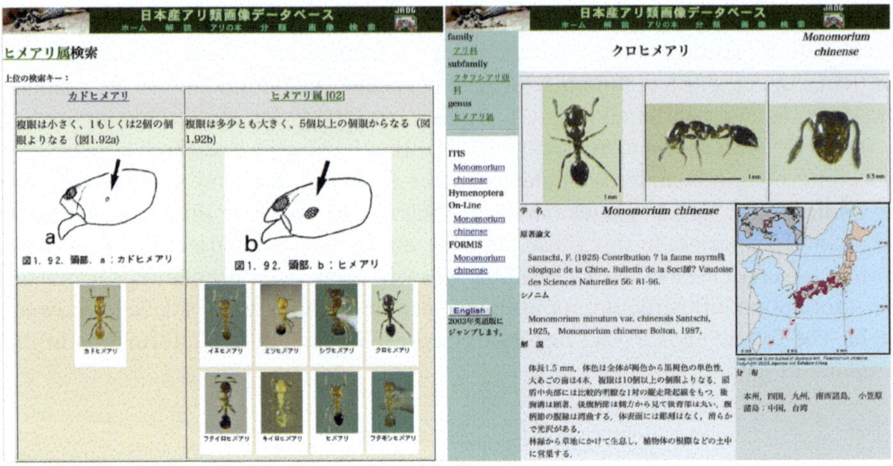


**Fig. 3.2** The website of the Japanese Ant Image Database

fundamental information infrastructures have strongly supported the development of ant research in Japan.

## 3.2.2 *Importance of the Accumulation and Curation of Specimens*

Although this might not be common knowledge, the accumulation of specimens is very important for natural science, and research designs with citizen-participation are no exception. For people other than specialists in taxonomy and related disciplines, many researchers are unfamiliar with specimens and their collections, and their utility is difficult to realize. However, in the case of taxa such as insects, which still have large gaps in taxonomic knowledge due to their enormous diversity, specimens are important evidence of the reliability of the data. As an example, in the case of the approximately 300 species of Japanese ants, the scientific names of more than 20% of these species have been changed since the publication of the 2008 edition of the Japanese Ant Image Database (Japanese Ant Database Group 2008 2008) mentioned above. These include not only the discovery of new species and new Japanese records but also those for which the combination of Japanese and scientific names has changed as a result of progress in overseas research, those that were thought to be a single species but have turned out to include multiple species, and even those that were thought to be multiple species but are in fact the same species. The impact of these name changes may be limited if the study design does not consider subsequent changes over time. However, one of the major points of information that local participants want to know in the context of citizen collaboration, such as information on the sustainable conservation of the local environment and the changes

**Fig. 3.3** Ant specimens accumulated at the research institute. Samples collected in surveys are sorted, identified, labeled, and then stored in the collection room

necessary to achieve this, is the comparison along this time axis. Even if the past observation records are not clear enough with only textual or image information, the authenticity and accuracy of the study can be verified if real samples with appropriate data are preserved. Such specimens can often show us evidence of phenomena such as regional extinctions of certain species due to land modification or environmental degradation. In the case of cooperative citizen surveys, citizen participation makes it possible to conduct surveys with more comprehensive coverage under unified protocol than what is possible with one researcher working alone. It is already valuable for all participants to establish a local specimen collection with accurate collection data. This is because it is nothing less than creating an asset, a biodiversity time capsule, for the area and its future (Fig. 3.3).

## 3.3 Practice and Achievements of Collaborative Study with Citizens Using Ants

### *3.3.1 Advantages and Challenges of Collaboration with Citizens on Ant Research*

From here, I would like to discuss the achievements and challenges based on our experience in designing and implementing a community-based collaborative biodiversity monitoring project using ants as target taxa. In the case of long-term projects such as the continuous biodiversity monitoring that we aimed for, it was important to establish a sustainable and cooperative relationship in a way that was beneficial to both parties, rather than have temporary participation. In the general citizen-science model, participating citizens provide materials and data collection, and researchers use them for analysis. Both sides benefit in this design; for example, the researcher gains contributed labor for the research through the cooperation of the citizens, and the citizens can participate in the research and enhance their knowledge and skills by gaining access to more specialized knowledge through the experience. Certainly, such cooperative relationships are beneficial to both the researcher and the citizen; at the moment, the research is underway. However, these common models, which the researcher is at the top of the structure, can be vulnerable regarding sustainability. Once the project stage moved to analyzing data aggregated from the initially designed survey model and publishing the research results, the collaboration system may become difficult to maintain because the cost-effectiveness of collaboration for the researcher side reduced, and their motivation weakened to maintain the relationships. This difficulty is also because if the contribution to the research from the citizen side is limited to the provision of data and materials based on the instructions of the researcher side, there is a lack of incentive from the citizen side to initiate a proactive bottom-up approach to research goals. Of course, this may also lead to ongoing and proactive efforts by the region based on the research results to solve that problem if the purpose and motivation of the research itself solve an environmental problem faced by the local community, for example. And the motivation of citizens to participate in proactive and active research greatly determines the sustainability of the collaborative system.

Another issue that needs to be addressed is the gap between the speed of research progress and the citizens' sense of waiting time. The collection of samples and data, especially in field science, is just the starting point in terms of research work from the researcher's point of view. Starting with the sorting of samples, identification, digitization of data, screening of input data, analysis, comparison with past studies, and constructing discussions, it may take years before the results can be open to the public. The time required for data screening for the analysis is expected to be even longer if the scale of the survey is larger due to the cooperation of citizens than was conducted by one researcher alone and if the survey skills and experience of the participants vary. On the other hand, the citizens participating in the survey have no choice but to wait for this period of time, even if they are informed and understand the schedule for these work processes. It will be necessary to consider measures to address these issues in order to realize mutually proactive collaboration between researchers and citizens when a citizen-participated research project is designed.

### 3.3.2 Bridging the Gap Between Citizens and Researchers

When we actually plan and implement citizen science, it is important for the planning side to design a system that assumes that there is a gap between the planning researcher and the participant citizens in terms of their motivation, willingness, and sense of purpose. From the point of view of researchers who plan citizen-initiated research, the goal is generally to collect data and use it to produce research results, but this goal is not necessarily the same as that of the citizens who will participate in the project. It would be ideal if researchers were involved in solving actual local issues, including social implementation, as in community science and action research, and if the goals were shared with all participants. The major motivation for providing labor from the citizen's perspective is that problem-solving for a local issue can be furthered by new scientific findings, which align with the researchers' personal goals. We should understand that it is not easy for the general public to truly grasp the perspective of researchers who focus on the quality of data and new scientific findings to be learned from the data. As a concrete example, I would like to introduce the case of our collaboration with a local high school on a survey of ants in Okinawa, Japan.

The goal of the research project was to establish a long-term biodiversity monitoring network in Okinawa prefecture, to observe and gain new insight into the current status and dynamics of the biota, and to invite high schools in the prefecture to participate in monitoring. In Okinawa, where nature and culture are closely interrelated, biodiversity conservation is of great concern to the people of the prefecture, and it is not too difficult to agree with the schools on the objectives of the research, which is to collect and accumulate basic data for this purpose in a uniform manner over a wide area. This is also helped by the fact that the northern Okinawa-Jima and Iriomote Islands have been registered as World Natural Heritage sites, and the control and monitoring of invasive alien species is one of the most urgent issues in their conservation management. On the other hand, however, the story suddenly becomes complicated when we consider how to involve high schools in the project. For the high school side, there is a clear educational benefit to expose students to research projects conducted by universities and the researchers who run them. However, the associated human and time costs for both students and teachers within the current school curriculum became a major issue. As far as ant surveys are concerned, studies on community dynamics generally involve random and nonselective collection methods such as pitfall or Malaise traps, or else the extraction of organisms from defined volumes of soil. Each sample collected from such traps or soil contains a large number of various organisms, including ants. The work necessary to maintain the traps and process the samples would far exceed the scope of work available for school classes or club activities if students were involved in all steps of the survey. In the end, high schools may be able to participate in an event-driven basis and experience small proportion of the total tasks, but it will be difficult for them to contribute to the research wholly and continuously to get a sense of the overall picture. On the high school side, on the other hand, they expect researchers to be involved in compact research themes that can be conducted independently by the students themselves within the curriculum for educational purposes. In addition, the differentiation of research goals among schools is also an important factor for them, as these high school research projects, as the inquiry-based learning, are to be exhibited in internal and external competitions. Unfortunately, the small and highly disparate datasets obtained through such high school independent research will be reduced considerably in the data-screening stage by researchers and may contribute little to the analysis. Although we use the same term, "research," there is a difference in meaning between the high school, where education is the top priority, and the research institution, where research is actually implemented for academic outputs (Fig. 3.4). In order to build a mutually beneficial relationship with institutions and organizations that have different objectives for establishment and operation, it is necessary to have a system that allows both parties to advance each other's

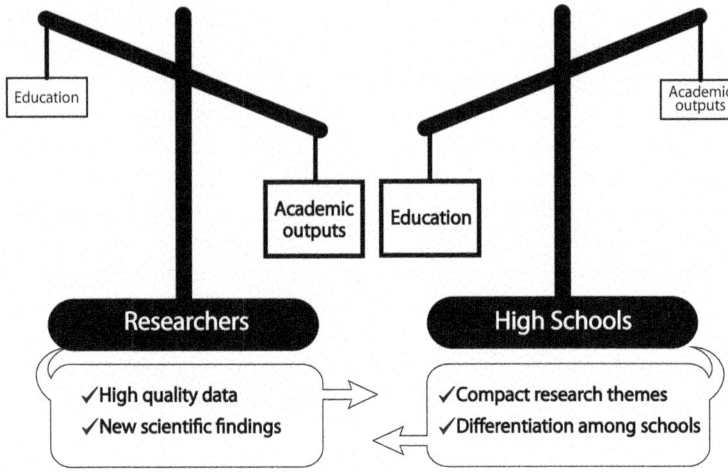


**Fig. 3.4** A gap in the purpose of research collaboration existing between the researchers and high schools. The researchers' side focuses on more academic output, while the high school side focuses on more education. Both sides have different expectations of each other (Illustrated by Mayuko Suwabe)

missions voluntarily and autonomously to a certain extent. Through such voluntary and autonomous efforts, intensive citizen science may evolve into action research and community science, which are more sustainable and have more real-world impact.

### *3.3.3 Development of a Survey Package for the Study of Ants in Okinawa in Conjunction with Local High School Exploration Activities*

We recognized the importance of improving problems caused by differences in objectives and skills between researchers and high schools, which are the main educational institutions, and improving sustainability of the system. We also recognized that it is difficult to use the simple one-way data-intensive system from the citizen to the researcher that has been used in the general citizen science framework. Therefore, we attempted a new strategy in which each participating organization, including the researchers, plans and implements projects on its own and connects them as a network. Specifically, each high school conducts its own research project while we carry out a prefecture-wide biodiversity monitoring project. This was achieved by linking these seemingly independent projects by sharing a unified ant survey protocol as the package, thereby sharing "survey results" through a network of projects. The development of this unified survey protocol was one key challenge to this networking project.

There are two major issues to be considered in a survey package that can be used for everything from high school research projects to full-scale, wide-area biodiversity monitoring. The first is the issue of survey cost. Each survey unit should finish approximately 30 min if the survey is to be conducted in a high school classroom. Furthermore, within the limited time of a single school year, it is necessary to obtain a volume of data that can be analyzed even on a small scale. Considering the need to conserve effort, we adopted the Time Unit Sampling Method. Although this method is relatively new and still in its infancy as an ant survey method compared to pitfalls, soil extraction, bait traps, and Malaise traps, its use is spreading, especially in Asia (e.g. Anh et al. 2010; Park et al. 2014; Rahman et al. 2022). Although this method takes much less time than the more common pitfall, soil extraction, or Malaise trap methods, from sample collection to sorting (Yoshimura et al. 2016), it still provides data in the form of species lists and their respective frequencies of occurrence. This sampling method is also less biased than baiting traps with honey, tuna, or other food item due to species-specific preferences for the food, and it is also less affected by the changes in activity of worker ants' foraging behavior due to changes in air temperature. Although there is a concern that the results may be more biased by the observers' collection skill and experience than with passive traps, it has been shown that this can be improved to some extent with prior practice and reminders (Ogata and Takematsu 2001). In addition to unitizing the collection time as the standard method, we also developed the method by unifying a single collection area to 400 square meters (mistakenly shown as 200 square meters in Yoshimura et al. (2016)). The 400-square-meter area can be set in a small urban park and coincides with the general forest survey quadrat area, making it easier to conduct comparisons in each survey section from the contiguous forest to the urban environment. Such a prepared packaged survey method is also helpful in reducing the effort of teachers responsible for providing multiple research topics to many students every year (Fig. 3.5).

The second point is the issue of data quality, especially relating to accuracy. Data compiled in high school research projects, which focus primarily on educational attainment goals, cannot be used for analysis by researchers without screening and modification because of issues with the accuracy of the species identification. Therefore, we ensured the quality of these data by asking the high schools to share not the data but the samples themselves, taking advantage of sharing the same survey protocol (Fig. 3.5). This sharing of materials reduces the burden of storing and managing samples on the high school side after their annual project is over, and it also can serve as the collection for future ant studies through re-identifying and re-organizing in the specimen curation system of the research institution. This system will allow us to separate the data processing between the high school side and the researcher side, even if we share the same specimens. This way, the goals and standards for quality of the data collected by participants will not affect each other. Basically, we provided the research methods, tools, and advice, and it was up to the schools to decide what topics they would like the students to use for their research. In addition to the above, we focused on working with biology teachers as mentors in schooling rather than with students, though working with students would be more

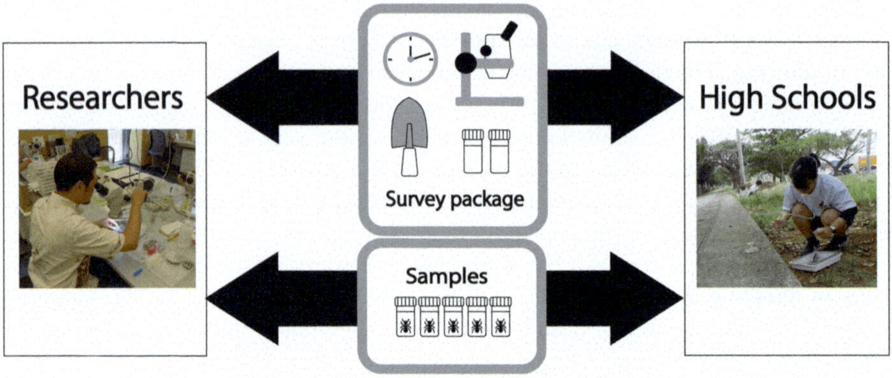


**Fig. 3.5** The sharing of easy-to-use survey packages and the samples collected using these methods allows for the coexistence of two research objectives: academic outputs and high school education. The sharing of the survey package is beneficial to the high schools in terms of cost-effectiveness and to the researchers in terms of ensuring sampling accuracy. Sharing the sample rather than the data allows for separate assurance of data quality for both parties' purposes (illustration: Mayuko Suwabe)

typical in an intensive citizen-science data collection. This is because the project team members in high schools change almost every year, making it difficult to accumulate and pass on knowledge and skills among the students. Our proposal allows the school to organize research independently and on an annual basis with minimal burden on the curriculum while researchers have access to samples collected according to a set protocol, creating a win-win situation. In fact, since 2015, when this collaborative project was launched, several schools in the prefecture have conducted some kind of research using this package, and some of them have successfully won prizes in prefectural competitions.

### *3.3.4 A Cross-Organizational Cooperative Effort to Take Countermeasures Toward Invasive Alien Ants*

In Okinawa where we are based, the importance of biodiversity monitoring has been widely recognized in recent years due to the registration of northern Okinawa-Jima and the Iriomote Islands as World Natural Heritage sites, as mentioned earlier, and the widely publicized issue of ecosystem disturbance by invasive alien species. In particular, the recent series of discoveries of invasive alien ants in or around port areas throughout Japan (Hashimoto 2020), including the first Japanese record of the Red Imported Fire Ant, *Solenopsis invicta* Buren, 1972, in Kobe in 2017, has led to a shared understanding that monitoring and control of ants is an essential and important issue for biodiversity conservation in Okinawa, as in Yoshimura et al. (2021). Under these circumstances, the biodiversity monitoring network we developed is evolving into an action research project in cooperation with the government and other related organizations as part of this effort. The development of the survey package for collaboration with the high school mentioned above has led to the adoption of similar protocol for the survey of invasive ants around the port area in collaboration with governmental agencies. This has enabled a rapid understanding of the current situation and tracking of the ant community dynamics in the port areas. After completing high school research projects, the ant samples provided by those schools are used to research the surveillance of invasive ants and their control as a part of the samples from the monitoring network. Through the development of the research material sharing system fueled by collaboration between local high schools and researchers, samples collected by the high school students themselves for their different research projects were reused to contribute to larger-scale biodiversity monitoring research.

### 3.3.5 *Future Improvements for Conducting a Collaborative Research Project Using Ants*

I think that our ant survey project with local high schools is a good example of how a collaborative project can bring more sustainability to citizen science, which has generally been researcher-driven and data-intensive. By sharing sample specimens but separating the sorting, databasing, and analyzing according to the achievement goals, we built a win-win system that allows two definitions of "ant research" to coexist: one defined at the academic level and one defined with school educational in mind. Since the high schools are also continuing their own projects, it can be said that this venture has achieved some success in terms of sustainability (Fig. 3.6).

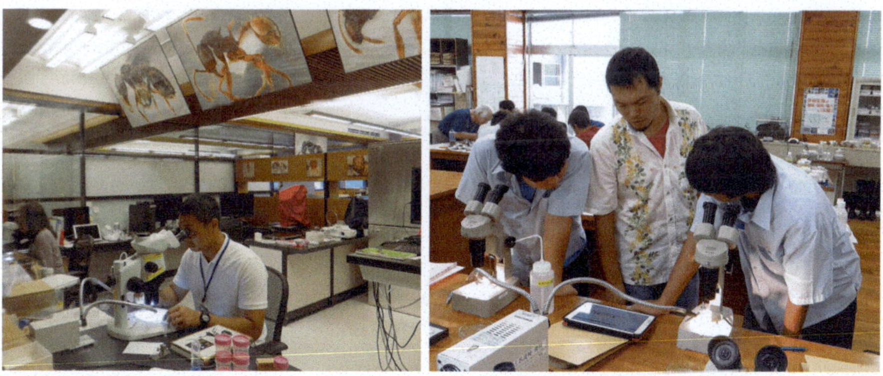

**Fig. 3.6** Interaction between research institutions and high schools. Left, the high school teachers visit the research institution to learn ant research techniques; right, the researchers occasionally visit high schools to provide expert advice to high school students and teachers

Several important issues have come to light in this process: two of which I would like to mention at the end of this chapter. One is the need to revise the implementation manual to include simpler identification tools. More than 5 years have passed since the start of collaboration with the high schools, and there have been significant changes due to the transfer of teachers. Some teachers are willing to launch new research projects at the new high schools they transferred to with their research experience using ants as research materials, and some have passed on the theme to their successors. The knowledge and skills of the instructor may be lost in the case that successors take over the project, and if the teachers themselves try to start up a new project at the new school, the data accumulation will start from scratch. In such cases, the species identification process is the costliest part of restarting a research project. While our team can assist with identifying the more difficult species, the students must work on more common species. The identification tools currently available cover about 300 species in all of Japan or about 150 species in all of Okinawa Prefecture, and keying out one species out of 150 or 300 is not so easy for students, who do not usually have any relevant experience. More compact and easy identification tools would thus be necessary. The ant fauna in and around high schools in the prefecture has been fairly well elucidated through the accumulation of data to date. For example, if we can develop an identification tool focusing on approximately 30 species in each area, identification work by high schools will become much easier. Search tables, simplified terminology, and picture matching using photographs of specimens would also be useful for educational purposes.

Second, we the researchers should provide data feedback to support the high school projects. As the participating high schools continue on their project year after year, they will accumulate data on the diversity of ants surrounding their school, and comparison and analysis of this data could be a theme in the next phase. The scope of the study at the high schools will be broadened if data from other areas of the prefecture accumulated under the same protocol can be used for this comparison and analysis. It may help raise the level of issue-based research in the prefecture if such data can be provided in a format that participating schools can handle. In the same way that our citizen science project developed into action research that enables researchers to contribute to solving local issues, it would be wonderful if participating schools could develop their own autonomous community science to make their activities more sustainable.

## References

Anh LN, Ogata K, Hosoishi S (2010) Ants of agricultural fields in Vietnam (Hymenoptera: Formicinae). Bull Inst Trop Agric Kyushu Univ 33:1–11. https://doi.org/10.11189/bita.33.1

Hashimoto Y (ed) (2020) Topics of alien ants. Asakura Publishing, Tokyo, 200 pp. [in Japanese]

Japanese Ant Database Group (1995) Japanese ant color image database. The Myrmecological Society of Japan. [CD-ROM], Mode of access: www.browser. [in Japanese]

Japanese Ant Database Group (1998) Japanese ant color image database. The Myrmecological Society of Japan. [CD-ROM], Mode of access: www.browser

Japanese Ant Database Group (2003a) Ant image database. Idengaku-Fukyukai (The association for propagation of the knowledge of genetics), Mishima. [CD-ROM], Mode of access: www.browser

Japanese Ant Database Group (2003b) Ants of Japan. Gakken, Tokyo, 224 pp

Japanese Ant Database Group (2003c) Super visual encyclopedia: ants of Japan. Gakken, Tokyo, 196 pp. [in Japanese]

Japanese Ant Database Group 2008 (2008) Japanese ant image database 2008. Japanese Ant Database Group, Sendai. [CD-ROM], Mode of access: www.browser. [in Japanese]

Myrmecological Society of Japan (1988) List of Japanese names of ants of Japan. Myrmecological Society of Japan, Tokyo, 59 pp. [in Japanese]

Myrmecological Society of Japan (1989) Identification guide and description of Japanese ants (I) Ponerinae, Cerapachyinae, Pseudomyrmecinae, Dorylinae, and Leptanillinae. Myrmecological Society of Japan, Tokyo, 53 pp. [in Japanese]

Myrmecological Society of Japan (1991) Identification guide and description of Japanese ants (II) Dolichoderinae and Formicinae. Myrmecological Society of Japan, Tokyo, 67 pp. [in Japanese]

Myrmecological Society of Japan (1992) Identification guide and description of Japanese ants (II) Myrmicinae and Leptanillinae (supplementation). Myrmecological Society of Japan, Tokyo, 105 pp. [in Japanese]

Ogata K, Takematsu Y (2001) II. Formicidae. In: Ogata K (ed) Study on factors affecting biodiversity monitoring. Grant-in-aid for scientific research report (C-2: 08680613), 7–27 pp. [in Japanese]

Park SH, Hosoishi S, Ogata K, Kuboki Y (2014) Clustering of ant communities and indicator species analysis using self-organizing maps. C R Biol 337:545–552. https://doi.org/10.1016/j.crvi.2014.07.003

Rahman MM, Sultana J, Haque Swapon MA, Ahamed T, Hassan J (2022) Diversity and functions of ants (Hymenoptera: Formicidae) in Bangabandhu Sheikh Mujibur Rahman Agricultural University Campus of Bangladesh. Trop Agroecosyst 3:27–33. https://doi.org/10.26480/taec.01.2022.27.33

Terayama M, Kihara A (1994) Distribution of Japanese ants by prefecture. Myrmecological Society of Japan, Tokyo, 73 pp. [in Japanese]

Terayama M, Takamine H, Kubota S (2009) Ants of Okinawa. Hidetsune Takamine, Naha City, 165 pp. [in Japanese]

Terayama M, Kubota S, Eguchi K (2014) Encyclopedia of Japanese ants. Asakura Shoten, Tokyo, 278 pp. [in Japanese]

Yamane S, Ikudome S, Terayama M (1999) Identification guide to the Aculeata of the Nansei Islands, Japan. Hokkaido University Press, Sapporo, 831 pp

Yamane S, Harada Y, Eguchi K (2010) Natural history of ants in South Kyushu, Japan. Nanposinsha, Kagoshima, 200 pp. [in Japanese]

Yoshimura M, Yoshida T, Ogasawara M, Economo EP (2016) Ant research program for high school students using the time-unit sampling method. Research Bulletin of Environmental Education Center, Miyagi University of Education 18:43–47. [in Japanese]

Yoshimura M, Ogasawara K, Suwabe M, Taga A, Naka M, Ogasawara M, Takehara Y, Tamashiro Y (2021) The first record of the Argentine ant *Linepithema humile* in Okinawa Prefecture (Hymenoptera: Formicidae: Dolichoderinae). The Biological Magazine Okinawa 59:1–9. [in Japanese]

Yoshizawa J, Mizoguchi T (2019) Identification guide book for familiar ants. Kasetsusha, Tokyo, 90 pp. [in Japanese]

# Chapter 4
# Hanamaru-Maruhana(Bumble Bee) National Census Using Photographs

**Yukari Suzuki-Ohno and Jun Yokoyama**

**Abstract** There are many online ecological community science programs that collect organisms' digital photographs taken by volunteers. Online ecological community science programs that collect digital photographs can be divided into two categories: those that collect photographs without specifying the target organisms (e.g., iNaturalist and iSpot) and those that collect photographs of the target organisms (e.g., eBird, Lost Ladybug Project, and Bumblebee Watch). In this chapter, we introduce our program using bumble bee photographs in Japan, the "*Hanamaru-Maruhana* national census." First, we describe the aims and procedure of this program. Second, we explain why we have used photographs to investigate the distributions of bumble bees. Target species influence the recording method and the types of volunteers participating in the program. Although bumble bees are not well known insects in Japan, more than 5000 bumble bee photographs were collected over 6 years in this program because photography was suitable for recording bumble bees. Third, we explain the outcome of this program, including scientific results obtained using the data collected in this program. Finally, we discuss the advantages and disadvantages of this program and the lessons learned from this program.

**Keywords** Citizen science · Deep learning · Invasive species · Occurrence data · Species distribution

Y. Suzuki-Ohno (✉)
Graduate School of Life Sciences, Tohoku University, Sendai, Miyagi, Japan

J. Yokoyama
Yamagata University, Yamagata-shi, Yamagata, Japan


Y. Suzuki-Ohno (ed.), *Community Science in Ecology*, Ecological Research Monographs, https://doi.org/10.1007/978-981-97-0304-3_4

## 4.1 What Is *Hanamaru-Maruhana* (Bumble Bee) National Census?

### 4.1.1 Aims and Procedure

The "*Hanamaru-Maruhana* national census (Bumble bee national census in English)" is an ecological community science program using photographs to investigate the distribution of bumble bees (*Bombus* spp.) in Japan (Fig. 4.1). Bumble bees are wild bees that are effective pollinators of both wild plants and cultivated crops. The conservation status of bumble bees is often addressed because the populations of some bumble bee species have been declining in Europe and North America (Williams and Osborne 2009; Cameron et al. 2011; Nieto et al. 2014; Kerr et al. 2015; Rasmont et al. 2015; Jacobson et al. 2018; Soroye et al. 2020). To gain a deeper understanding of bumble bees in Japan, it is necessary to investigate the country-wide distribution of bumble bees. In 2013, professional scientists from Tohoku University and Yamagata University launched the ecological community science program, the "*Hanamaru-Maruhana* national census." The aims of the "*Hanamaru-Maruhana* national census" are as follows:

I. To establish a method for collecting biodiversity data from volunteers.
II. To enhance volunteers' interest in bumble bee conservation.
III. To collect bumble bee data and estimate their ranges through species distribution models.
IV. To estimate the range shifts of bumble bees in relation to land use and global climate changes.
V. To develop bumble bee conservation plans based on bumble bee data and estimated ranges.

The main professional scientists of the *Hanamaru-Maruhana* national census asked volunteers to take bee photographs and send them via e-mail (Fig. 4.2) using web pages (https://hanamaruproject.s1009.xrea.com/hanamaru_project/index_E.html), Facebook, X (formerly Twitter), and flyers. Volunteers could participate in the *Hanamaru-Maruhana* national census without registering; however, they needed to register if they wanted to be included in the rankings and join mailing lists. Volunteers could send bee photographs to two e-mail addresses: the e-mail address of this program or the e-mail address of "Mobile Phone System and Cloud Services" provided by Fujitsu FIP, Co. The "Mobile Phone System and Cloud Services" are photograph database tools developed by Fujitsu, which enabled us to view the photographs, visualize where the photographs were taken using a web mapping platform, and download the occurrence data as a csv file. However, since the "Mobile Phone System and Cloud Services" are currently unavailable, we receive bee photographs via only the e-mail address of this program. The latitude-longitude data of the observation sites were extracted from the GPS data in the Exif information of the photographs. Although approximately 40% of photographs sent by volunteers had no GPS data, the latitude-longitude data or the addresses of the observation sites

**Fig. 4.1** Sixteen bumble bee species inhabiting the Japanese archipelago excluding the Kurile Islands (after Suzuki-Ohno et al. 2018). (**a**) *Bombus diversus* (Photographer: H. Morishima); (**b**) *B. ardens ardens* (Photographer: H. Morishima); (**c**) *B. hypocrita* (Photographer: H. Morishima); (**d**) *B. ignitus* (Photographer: Myano); (**e**) *B. honshuensis* (Photographer: H. Morishima); (**f**) *B. beaticola* (Photographer: K. Minemura); (**g**) *B. pseudobaicalensis* (Photographer: Y. Ishimori); (**h**) *B. deuteronymus* (Photographer: K. Minemura); (**i**) *B. ussurensis* (Photographer: K. Minemura); (**j**) *B. consobrinus* (Photographer: K. Minemura); (**k**) *B. norvegicus* (Photographer: K. Minemura); (**l**) *B. cryptarum* (Photographer: T. Tachibana); (**m**) *B. hypnorum* (Photographer: Y. Ishimori); (**n**) *B. schrencki* (Photographer: Y. Shirosaka (Hirabayashi), Copyright belongs to the Bihoro museum); (**o**) *B. yezoensis* (Photographer: Y. Ishimori); and (**p**) *B. terrestris* (Photographer: Y. Ishimori). (**a**)–(**o**) are native species. Subspecies of (**h**) *B. deuteronymus* (*B. deuteronymus maruhanabachi*), (**i**) *B. ussurensis*, (**j**) *B. consobrinus*, (**k**) *B. norvegicus*, and (**l**) *B. cryptarum* are rare species defined in this program. (**l**)–(**o**) were observed in only the northern island, Hokkaido. (**p**) *B. terrestris* is alien species

were found in the text in an e-mail (Suzuki-Ohno et al. 2017). The address was converted to latitude-longitude data using a web mapping platform (Google Maps). The species of bee photographs were identified from photographic images by Jun Yokoyama, who is one of the main professional scientists in the program and authors of this chapter.

For the success of a program, it is important to understand what kind of organism the target is. In the next subsection, we describe how the target species influences the recording method and the types of volunteers. In addition, we explain the details of the target species, bumble bees in Japan, in the Appendix.

### 4.1.2 Target Species, Recording Method, and Types of Volunteers

In general, it is difficult for nonexperts to identify bee species (Fig. 4.1, Appendix). Due to the difficulty in identifying bumble bee species, we cannot depend on eyewitnesses from volunteers. Although the specimens enable us to identify species correctly, the specimens have the risk of hurting volunteers or killing rare bee species. Therefore, we collected bee photographs taken by volunteers. It was relatively easy for experts to identify Japanese bumble bee species from photographic images because there are only 16 bumble bee species that inhabit the Japanese archipelago excluding the Kurile Islands (Appendix). It was much fewer than those in Europe and North America. The consistency of bumble bee species identification by Yokoyama (one of the main professional scientists in the program and authors of this chapter) was 95% for 15 bumble bee species and 97.7% for six major bumble bee species (Suzuki-Ohno et al. 2017).

Organisms that are suitable for photography are immobile or slow-moving species observed at a low position. Fortunately, bumble bees are slow and relatively insensitive to observers when they visit flowers. The program provided volunteers with information about bumble bees' favorite flowers of hedges and grasslands

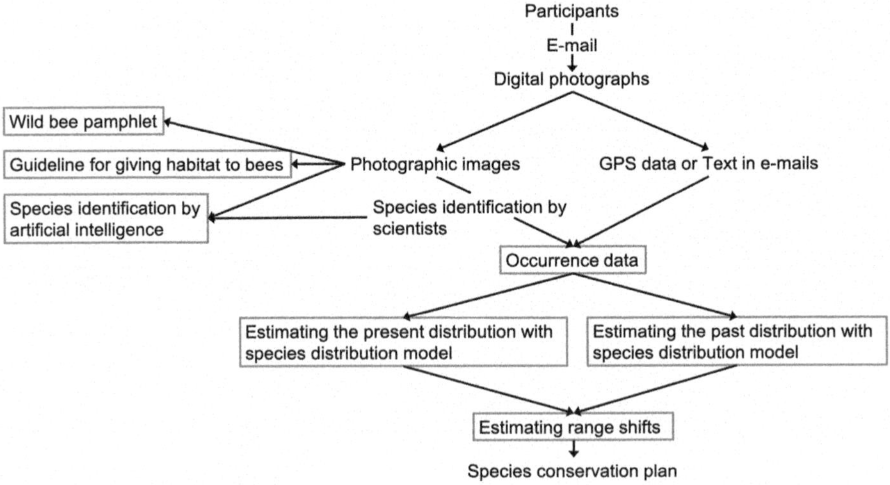


**Fig. 4.2** Outline of this program

(Fig. 4.3) through direct contact via e-mail or our web page. Volunteers found these flowers and waited to take photographs of the bumble bees. Almost all the bumble bee photographs sent by volunteers were bumble bees collecting nectar and pollen from the flowers.

Digital photography is a widely accepted method for volunteers to record organisms in community science programs. Participants in the *Hanamaru-Maruhana* national census include not only those interested in biodiversity conservation but also those who just like insects, flowers, and gardening; those who grow crops pollinated by bees; or those who have a photography hobby. Not all the volunteers had detailed knowledge about bees; however, they could equally contribute to the program by taking and sending bee photographs.

**Fig. 4.3** Favorite flowers for bumble bees. We selected familiar flowers that can be observed by volunteers. (**a**) *Styrax japonica* (Photographer: H. Morishima), (**b**) *Enkianthus perulatus* (Photographer: Y. Suzuki-Ohno), (**c**) *Rhododendron* sp. (Photographer: Y. Suzuki-Ohno), (**d**) *Cirsium* sp. (Photographer: H. Morishima), (**e**) *Impatiens textorii* (Photographer: H. Morishima), (**f**) *Aconitum* sp. (Photographer: H. Morishima), (**g**) *Trifolium* sp. (Photographer: Y. Suzuki-Ohno), (**h**) *Abelia* × *grandiflora* (Photographer: H. Morishima), and (**i**) *Cosmos bipinnatus* (Photographer: H. Morishima)

## 4.2 The Outcome of *Hanamaru-Maruhana* National Census

Approximately 1000 bee photographs are collected annually through the program (Table 4.1). These bee occurrence data from 2006 to 2018 (Fig. 4.4) are available in the open databases of the Global Biodiversity Information Facility (GBIF) and Japan Long-term Ecological Research Network (JaLTER) (Suzuki-Ohno et al. 2021a). Occurrence data from 2013 to 2015 were used to estimate the invasion risk of *B. ignitus* to non-native areas (Ministry of the Environment, and Ministry of Agriculture, Forestry and Fisheries 2017; Suzuki-Ohno et al. 2017) and the present and past distributions of the six major bumble bee species in Japan (Figs. 4.2, 4.5, and 4.6; Suzuki-Ohno et al. 2017, 2020a). Bee photographs were used to train a deep learning model that can identify bee species (Figs. 4.2 and 4.7; Suzuki-Ohno et al. 2022; BeeMachine https://beemachine.ai) and to create an educational pamphlet about bees (Figs. 4.2 and 4.8; Suzuki-Ohno et al. 2020b) and guidelines for giving habitat to bees (Figs. 4.2 and 4.9; Suzuki-Ohno et al. 2021b). Based on the results above, we believe that the program has achieved aims I, III, and IV. We continue the program for achieving aims II and V.

Here, we introduce the outcome of the program, including scientific research and other activities in the following subsections: "Bee Occurrence Data" (Suzuki-Ohno et al. 2021a), "Estimation of the Invasion Risk of *Bombus ignitus* to Non-native Areas, Hokkaido and Okinawa" (Ministry of the Environment, and Ministry of Agriculture, Forestry and Fisheries 2017; Suzuki-Ohno et al. 2017), "Estimation of Present and Past Distributions" (Suzuki-Ohno et al. 2017, 2020a), "Species Identification of Bee Photographs by Deep Learning" (Suzuki-Ohno et al. 2022), and "Wild Bee Pamphlet and Guidelines for Giving Habitat to Bees" (Suzuki-Ohno et al. 2020b, 2021b).

### *4.2.1 Bee Occurrence Data (Suzuki-Ohno et al. 2021a)*

Based on photographs collected from the *Hanamaru-Maruhana* national census, we published the occurrence data of 16 bumble bee species and three carpenter bee species in open databases (GBIF: https://www.gbif.org/dataset/5f7aeace-46b9-4516-8b1e-09650aee9024, and JaLTER: https://db.cger.nies.go.jp/JaLTER/metacat/metacat/ERDP-2021-04.1/jalter-en). We collected 5503 and 380 records of bumble bees and carpenter bees, respectively (Table 4.1 and Fig. 4.4). Although carpenter bees were not the target species, most of them were collected because of their similarity to bumble bees.

The occurrence data of bumble bees and carpenter bees cover a wide area in Japan. The location range of the occurrence data covers a latitudinal range of 26.45–45.29 and a longitudinal range of 127.80–145.82. The elevational range of the occurrence data was also wide and included high elevations (i.e., more than 1000 m). Ecological community science data usually tend to be skewed to

**Table 4.1** The number of records in bee occurrence data. The bee occurrence data included duplicate records such as several photographs taken at the same site on the same date (after Suzuki-Ohno et al. 2021a)

| | –2012 | 2013 | 2014 | 2015 | 2016 | 2017 | 2018 | **Total** |
|---|---|---|---|---|---|---|---|---|
| **Bumble bee species** | | | | | | | | |
| *Bombus diversus* [a] | 22 | 351 | 298 | 229 | 184 | 89 | 240 | 1413 |
| *B. ardens* [a] | 23 | 262 | 316 | 218 | 104 | 135 | 342 | 1400 |
| *B. hypocrita* [a] | 17 | 119 | 156 | 74 | 109 | 102 | 116 | 693 |
| *B. honshuensis* [a] | 1 | 47 | 94 | 65 | 37 | 34 | 7 | 285 |
| *B. beaticola* [a] | 5 | 38 | 61 | 41 | 13 | 29 | 10 | 197 |
| *B. hypnorum koropokkrus* | 7 | 13 | 3 | 7 | 12 | 33 | 59 | 134 |
| *B. deuteronymus deuteronymus*, *B. pseudobaicalensis* [b] | 1 | 4 | 2 | 19 | 40 | 3 | 17 | 86 |
| *B. yezoensis* | 1 | 8 | 0 | 2 | 7 | 8 | 6 | 32 |
| *B. schrencki albidopleuralis* | 2 | 4 | 2 | 0 | 3 | 2 | 7 | 20 |
| *B. terrestris* (alien species) | 7 | 18 | 4 | 69 | 44 | 38 | 65 | 245 |
| **Rare bumble bee species/subspecies** | | | | | | | | |
| *B. ignitus* [c] | 2 | 125 | 121 | 40 | 36 | 37 | 57 | 418 |
| *B. deuteronymus maruhanabachi* [c] | 0 | 4 | 1 | 69 | 127 | 56 | 40 | 297 |
| *B. consobrinus wittenburgi* [c] | 0 | 1 | 51 | 79 | 36 | 26 | 0 | 193 |
| *B. ussurensis* [c] | 0 | 0 | 3 | 34 | 18 | 12 | 4 | 71 |
| *B. cryptarum* [c] | 0 | 0 | 0 | 0 | 10 | 0 | 2 | 12 |
| *B. norvegicus japonicus*[c] (kleptoparasitic species) | 0 | 0 | 0 | 7 | 0 | 0 | 0 | 7 |
| **Carpenter bee species** | | | | | | | | |
| *Xylocopa appendiculata circumvolans* | 11 | 202 | 79 | 23 | 22 | 18 | 20 | 375 |
| *X. flavifrons* | 0 | 0 | 0 | 0 | 3 | 0 | 0 | 3 |
| *X. tranquebarorum* (alien species) | 0 | 2 | 0 | 0 | 0 | 0 | 0 | 2 |
| **Total** | 99 | 1198 | 1191 | 976 | 805 | 622 | 992 | 5883 |

[a] The occurrence data of these species include all subspecies in the Japanese archipelago excluding the Kurile Islands ("*Bombus diversus*" includes *B. diversus diversus* and *B. diversus tersatus*. "*Bombus ardens*" includes *B. ardens ardens*, *B. ardens sakagamii*, and *B. ardens tsushimanus*. "*Bombus hypocrita*" includes *B. hypocrita hypocrita* and *B. hypocrita sapporoensis*. "*Bombus honshuensis*" includes *B. honshuensis honshuensis* and *B. honshuensis tkalcui*. "*Bombus beaticola*" includes *B. beaticola beaticola* and *B. beaticola moshkarareppus*). *Bombus diversus tersatus*, *B. ardens sakagamii*, *B. hypocrita sapporoensis*, *B. honshuensis tkalcui*, and *B. beaticola moshkarareppus* inhabit Hokkaido. *Bombus ardens tsushimanus* inhabit Tsushima

[b] The sum of the photographs of *B. deuteronymus deuteronymus* and *B. pseudobaicalensis* was described because it was too difficult to discriminate them using photographic images alone (their difference is only a bunch of black hair on the second segment of the abdomen)

[c] These species are rare bumble bee species/subspecies defined in Suzuki-Ohno et al. (2021a); however, *B. ignitus* is common in some regions of Japan. The latitude-longitude data of these six rare species/subspecies were rounded off to one decimal place to protect their habitats

low-elevation areas such as urban areas. However, in the *Hanamaru-Maruhana* national census, some volunteers sent bumble bee photographs taken during mountaineering or staying in highlands (e.g., summer resorts). The date range of

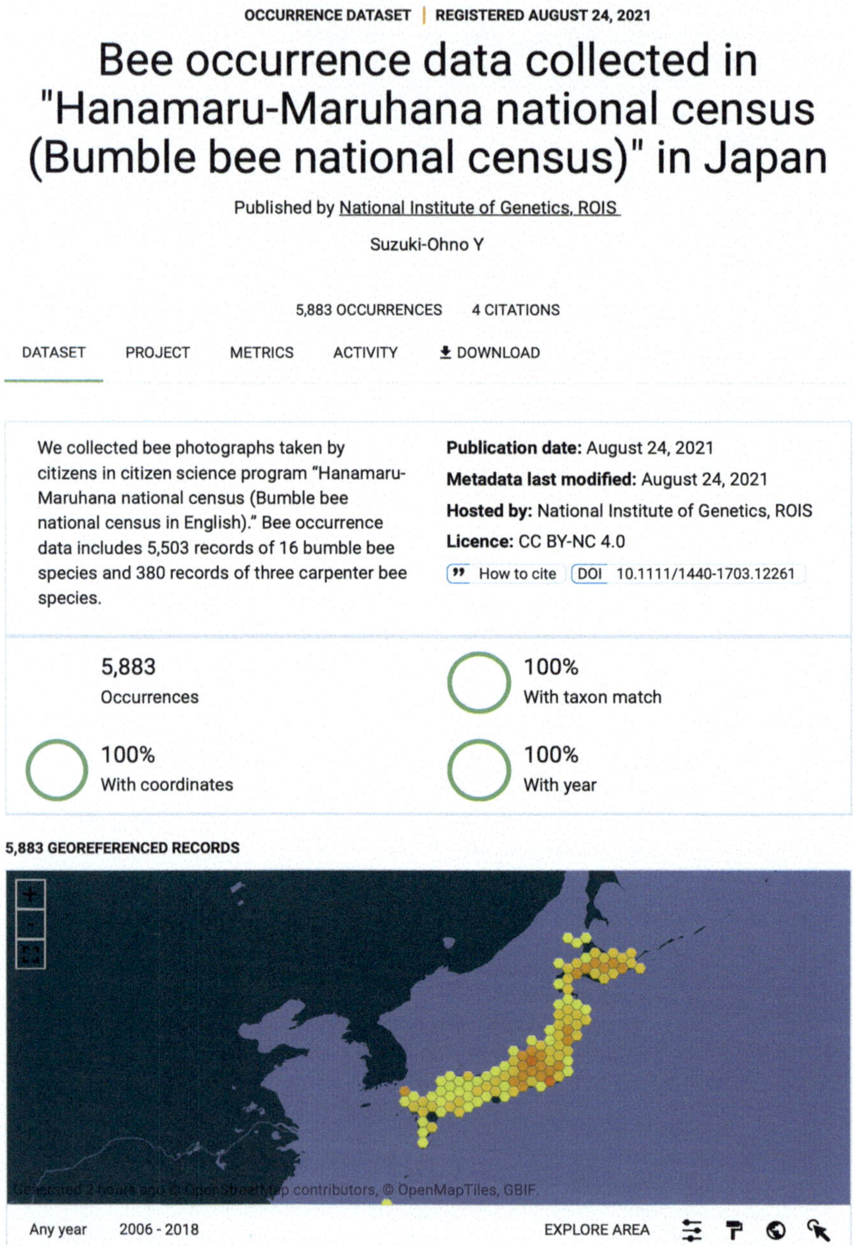


**Fig. 4.4** Snapshot of the website of bee occurrence data obtained from this program in Global Biodiversity Information Facility (GBIF). https://www.gbif.org/dataset/5f7aeace-46b9-4516-8b1e-09650aee9024. Accessed 11 Jan 2023

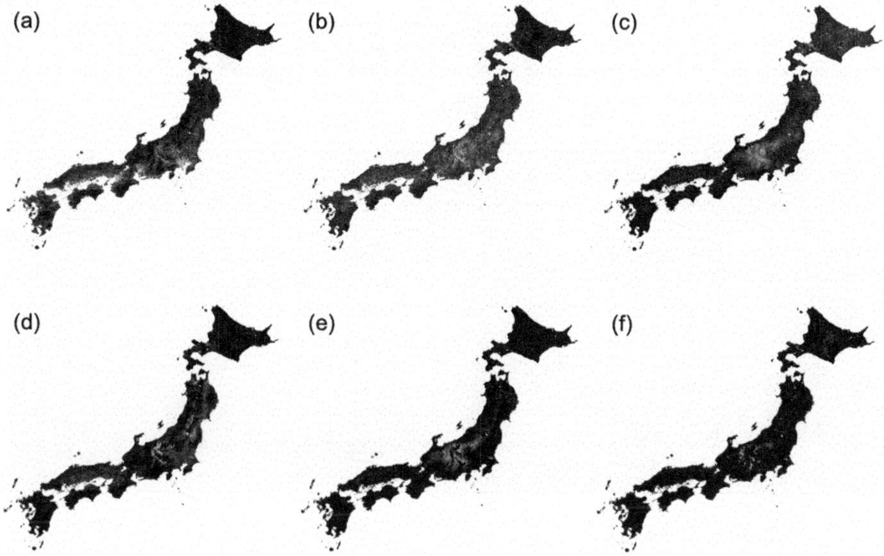


**Fig. 4.5** Estimated distributions of (**a**) *Bombus ardens*, (**b**) *B. diversus*, (**c**) *B. hypocrita*, (**d**) *B. ignitus*, (**e**) *B. honshuensis*, and (**f**) *B. beaticola* (after Suzuki-Ohno et al. 2017). White represents a high probability of distribution area whereas black represents a low probability of distribution area. This map was generated with the software ArcGIS ver. 10.0 (https://www.arcgis.com/features/index.html)

**Fig. 4.6** Estimated possible range shifts of (**a**) *Bombus hypocrita* and (**b**) *B. beaticola* (after Suzuki-Ohno et al. 2020a). Black represents the estimated range reduction. Gray and white represent the estimated unchanged range and range expansion, respectively. This map was generated with the software QGIS ver. 2.14 (https://qgis.org/en/site/)

occurrence data covers September 2006 to October 2018. Although the program started in 2013, volunteers kindly sent bee photographs taken before the program started. Part of this occurrence data (from 2013 to 2015) was used for the bumble bee distribution estimation studies explained in the following two subsections. The

**Fig. 4.7** Screenshot of the website “Bee Machine.” https://beemachine.ai. Accessed 11 Jan 2023

**Fig. 4.8** Wild bee pamphlet (Suzuki-Ohno et al. 2020b). Cover page (left) and the first page (right)

## 2. 野生ハナバチとは？

### 単独性ハナバチ

野生のハナバチは、一般的には1cm程度、小さいものは5mmほどから、大きいものは2cm以上になります。ハナバチのほとんどの種類が単独性といって、雌蜂が一匹で、姉妹や働き蜂の手助けなしに自分の巣を作り、幼虫の餌を用意します。みなさんがよく見かける単独性ハナバチは、ハキリバチ類やヒゲナガハナバチ類、クマバチ類などです（図1）。キムネクマバチは、雌蜂が一匹で子育てしますが、娘が同じ巣で生活する亜社会性を持ちます。

### 単独性ハナバチの巣

野生ハナバチの多くが、地中に巣を作ります。ほとんどの場合、雌蜂が一匹で、自分の巣のトンネルを掘り、このトンネルにいくつかの部屋（育房）を作ります。他に、竹筒や空洞の茎のようなものの中に巣を作るものもいれば、甲虫の幼虫が木材にあけた穴に巣を作るもの、自ら木材に穴をあけて巣を作るものもいます。こういった竹筒や木に営巣する雌蜂は、葉の欠片や泥などの材料を仕切りに使って、一列に並んだ育房を作ります。育房には、孵化した幼虫のために花粉と蜜を混ぜた団子を置いて、卵を産みます。

単独性ハナバチの巣には、労働寄生のハナバチが寄生することがあります（付録A）。

### 社会性ハナバチ

よく見かける野生の社会性ハナバチは、ミツバチとマルハナバチです（図2）。日本のミツバチはニホンミツバチの1種だけですが、マルハナバチは15種生息しています。社会性ハナバチは、女王蜂だけが卵を産み、一つのコロニー（巣に棲む集団）に多くの働き蜂が育ちます。ミツバチは働き蜂が数百～数万、マルハナバチは数十～数百と言われています。

蜂は針で刺すイメージがあると思いますが、ハナバチの中で刺すとすれば、だいたいミツバチかマルハナバチになります。ただ、自分への危険や、巣への危険を感じなければ刺すことはなく、花に訪れて蜜や花粉を集めているハナバチを必要以上に怖がる必要はありません。

図2：日本に生息する野生の社会性ハナバチの例。上からニホンミツバチ、コマルハナバチ、トラマルハナバチ。

図3：木の洞に作られたニホンミツバチの巣。

**Fig. 4.9** The guidelines for giving habitat to bees (Suzuki-Ohno et al. 2021b). One of the pages explains solitary bees and social bees

*Hanamaru-Maruhana* national census is still ongoing, so occurrence data will continue to increase.

### *4.2.2 Estimation of the Invasion Risk of* **Bombus ignitus** *to Non-native Areas, Hokkaido and Okinawa (Ministry of the Environment, and Ministry of Agriculture, Forestry and Fisheries 2017; Suzuki-Ohno et al. 2017)*

In Hokkaido, the invasion of the alien bumble bee species *B. terrestris* has been reported to have negative impacts on native species inhabiting Hokkaido (Inoue et al. 2008; Inoue and Yokoyama 2010). The import of commercial *B. terrestris* colonies began in approximately 1992, and their wild nests were found in Hokkaido in 1996. Commercial colonies have been used for the pollination of tomatoes in greenhouses. As an alternative to *B. terrestris*, the use of *B. ignitus* was considered because *B. ignitus* was the only native species sold as commercial colonies. Commercial colonies of *B. ignitus* have been used in Honshu, Shikoku, Kyushu, and southern islands such as Okinawa, but not in Hokkaido. Although *B. ignitus* is a native species, it does not inhabit Hokkaido and Okinawa. *Bombus ignitus* may successfully invade Hokkaido as a domestic alien species and have negative impacts on native species inhabiting Hokkaido. Before commercial colonies of *B. ignitus* are used, it will be necessary to estimate the invasive potential of *B. ignitus* in Hokkaido. Therefore, we estimated the possibility of *B. ignitus* invasion in Hokkaido and Okinawa using occurrence data from the *Hanamaru-Maruhana* national census and environmental data with a species distribution model, Maxent (Phillips et al. 2006). The estimations showed that the possibility of invasion into urban areas at low elevations in Hokkaido was sufficiently high, whereas the possibility of invasion into Okinawa was low (Suzuki-Ohno et al. 2017). M. Kawata and J. Yokoyama, the two professional scientists of the *Hanamaru-Maruhana* national census, participated in the committee on *B. ignitus* use in the Ministry of the Environment, and the results of this study contributed to policy formulation. The Ministry of the Environment and the Ministry of Fisheries announced a policy not to introduce commercial colonies of *B. ignitus* in Hokkaido (Ministry of the Environment, and Ministry of Agriculture, Forestry and Fisheries 2017).

### *4.2.3 Estimation of Present and Past Distributions (Suzuki-Ohno et al. 2017, 2020a)*

The causes of bumble bee declines have been considered to be several factors or their combinations: global warming, land-use change, pesticides, and diseases, including parasites. In the *Hanamaru-Maruhana* national census, we focused on bumble bee range shifts caused by global warming and land-use changes. First, we estimated the present distributions of six bumble bee species using current bumble bee occurrence data and present climate and land cover data with Maxent (Suzuki-Ohno et al. 2017; Fig. 4.5). Then, we estimated the past/present distributions of six bumble bee species using current bumble bee occurrence data and past/present climate and land cover data with Maxent (Suzuki-Ohno et al. 2020a; Fig. 4.6). We compared the estimated past and present distributions of each species and identified the differences between the two as possible range shifts. The estimated possible range shifts suggested that changes in climate and land cover shrank the estimated ranges of *B. diversus*, *B. hypocrita*, *B. ignitus*, *B. honshuensis*, and *B. beaticola* but expanded the estimated range of *B. ardens*. The estimated *B. hypocrita* and *B. beaticola* range reductions were severe in Hokkaido, the northern large island of the Japanese archipelago (Fig. 4.6). These estimated range reductions were mainly attributed to the increase in temperature.

According to the results of Maxent, the range reduction of *B. diversus* was partially attributed to large-scale increases in forest areas in local regions. Such increases in forest area may result from the abandonment of agricultural lands and the extension of the rotation time of planted coniferous forests or secondary forests. Abandoned planted coniferous forests have few understory flowering plants because of a lack of sunlight on the forest floor, which may be unsuitable habitats for bumble bees. Managing old planted coniferous forests and secondary forests will be the key to bumble bee conservation within the context of adapting to climate change.

### *4.2.4 Species Identification of Bee Photographs by Deep Learning (Suzuki-Ohno et al. 2022)*

Species identification of bees by nonexperts is often difficult because of the interspecific similarity and intraspecific variation of bees. Species identification of bees by experts is desirable in ecological community science programs, and the results of species identification by experts provide several benefits to the participants in the programs. However, species identification by a small number of experts is time- and effort-consuming, and the results of species identification by experts are not always available to the participants immediately. Therefore, we focused on deep learning, which has shown remarkable developments in image analysis. A deep convolutional neural network, Xception (Chollet 2017), was selected to identify bee species based on deep learning. Transfer learning (Oquab et al. 2014; Yosinski et al. 2014) and data augmentation (Perez and Wang 2017) have been adopted to resolve the issue of a shortage of photographs. A total of 3779 photographs of ten bumble bee species and two honey bee species collected in the *Hanamaru-Maruhana* national census (Suzuki-Ohno et al. 2017) were used as the training data for Xception. Xception learned species classes of the photograph dataset that were classified into different classes depending on the bee species (hereafter referred to as the species class experiment). Xception also learned color classes of the photograph dataset that were classified into different classes depending on color differences resulting from intraspecific variation, in addition to bee species (hereafter referred to as the color class experiment).

The total accuracy of species identification by Xception reached 83.4% versus an expert's identification in the species class experiment. Moreover, the total accuracy of species identification by Xception increased to 84.7% versus an expert's identification in the color class experiment. Both the negative effects of intraspecific variation and interspecific similarity were mitigated in the color class experiment. These accuracies (83.4% and 84.7%) were much higher than the accuracy of species identification by 50 biologists (53.7%).

We also cooperated with Spiesman's worldwide project, BeeMachine (Fig. 4.7; https://beemachine.ai). The project aims to identify the species of bumble bee photographs taken by people around the world. We have provided the project with the bumble bee photographs collected in the *Hanamaru-Maruhana* national census.

### *4.2.5 Wild Bee Pamphlet and Guidelines for Giving Habitat to Bees (Suzuki-Ohno et al. 2020b, 2021b)*

Yukari Suzuki-Ohno and Jun Yokoyama, the two professional scientists of the *Hanamaru-Maruhana* national census, belong to the Pollination Services Society of Japan (PSSJ). Photographs provided by the *Hanamaru-Maruhana* national census were used in a wild bee pamphlet for children (Fig. 4.8; Suzuki-Ohno et al. 2020b). Photographs provided by the *Hanamaru-Maruhana* national census were also used in the conservation guidelines of the PSSJ (Fig. 4.9; Suzuki-Ohno et al. 2021b). Before making the guidelines, some members of the PSSJ translated the guidelines, "Farming for bees," published by the Xerces Society (https://www.xerces.org/publications/guidelines/farming-for-bees) into Japanese (https://sites.google.com/site/pollinationservicessociety/home/farming-for-bees-in-japanese). Using this as a reference, we recreated the guidelines for giving habitat to bees, paying attention to the differences between the United States of America and Japan. These are available on the website of the PSSJ (https://sites.google.com/site/pollinationservicessociety/home/guideline).

## 4.3 The Advantages, Disadvantages, and Lessons Learned from the *Hanamaru-Maruhana* National Census

We believe that this program was successful because aims I, III, and IV were achieved by collecting many bee photographs and conducting scientific research using these photographs. In addition, we could publish a wild bee pamphlet and guidelines for giving habitat to bees. However, every program has its advantages and disadvantages, and in some cases, advantages can become disadvantages. In this section, we summarize the advantages, disadvantages, and lessons learned from the program.

## *4.3.1 The Advantages of this Program*

### 4.3.1.1 Versatility of Digital Photographs

Digital photographs contain a variety of information such as bumble bee images, background images, and GPS data. With the bumble bee images, it was possible for experts to identify the species. Photographs are easy to share and can increase the accuracy of species identification through checking by multiple experts. Background images sometimes help to identify bumble bee species. For example, the presence of certain flowers in the background can make species identification easier because bumble bees' favorite flowers and flower-visiting behavior often differ among bumble bee species. Thanks to GPS data from photographs, we can obtain the exact latitude and longitude data of the observation sites. This is an important point when collecting presence data. In addition to presence data, we can consider various uses of bumble bee photographs, such as being training data for a deep learning model, data of phenotypic variation, and data of visiting flowers. This is an advantage that leads to the collection of accurate data and the application of other studies.

### 4.3.1.2 Simplicity and Ease of Participation Method

This program is an online community science program that collects bee photographs via e-mail without app installation or registration. In addition, the participants did not need to identify the species in their photographs. The simplicity and ease of the participation method allow for an unspecified number of volunteers to participate as much as they like at any time and place. They could easily send many bee photographs to the program. This is an advantage that leads to the collection of occurrence data over a wide area, including places that are difficult for individual scientists to survey, such as private gardens and remote islands.

### 4.3.1.3 Target Species, Bumble Bees

Bumble bees are easy to find and observe because they are relatively large and diurnal. Since bumble bees are important pollinators and conservation target species, it is easy to communicate the importance of the survey in this program. Even people who are not interested in bees have participated in this program because bumble bees are important pollinators for plants, including vegetables and fruits.

The number of target bumble bee species in Japan was good for this program. Too much will make species identification difficult and too little will make photography difficult in some places. Target bumble bee species include common species and rare species, and their variation has a positive effect on maintaining the motivation of the participants. Common bumble bee species are relatively easy to observe when they collect nectar and pollen from flower beds, hedges, and flowers of crops. In addition, we could mitigate the effect of sampling bias using the total occurrence data from all species (Suzuki-Ohno et al. 2017). This is an advantage that leads to the estimation of bumble bee distributions.

It was also an advantage that there were many previous studies on bumble bees and their foraging range. Although bumble bees can fly over several kilometers, they often forage at approximately 300 m from their nests (Osborne et al. 1999). In addition, it has been reported that land use within a circle with a 750 m radius affects the nesting sites of bumble bees (Steffan-Dewenter et al. 2002). Therefore, we set the error range of the observation sites to less than 500 m.

## 4.3.2 *The Disadvantages of this Program*

### 4.3.2.1 Time and Effort of Species Identification by Experts

In this program, species identification in photographs did not have to be performed by the participants. The species in the photographs had to be identified by a small number of experts. Species identification by a small number of experts is time- and effort-consuming. It also caused difficulty in responding to the requests of participants who wanted to know the species name immediately. To reduce the time and effort of species identification by experts, it is possible to identify species by the participants or by using deep learning. Species identification by the participants or using deep learning will be useful for participants and experts. However, verification by experts is usually required when photographs are used as the presence data at the species level for scientific research or conservation activities. The time and effort of species identification by experts will be difficult to reduce if photographs are used as the presence data at the species level.

### 4.3.2.2 Communication Via E-mail

We selected e-mail to collect photographs, which made it unnecessary to meet directly with the participants and impossible to share the experience of field research. It tends to be difficult to enhance interest in nature and bumble bees.

In addition, e-mail made it difficult to confirm the transfer of photograph copyrights. Although it was written on the web page that the copyright of the photographs would be transferred to the *Hanamaru-Maruhana* national census at the time of posting, some participants did not read it. Since we aimed to collect occurrence data, we tried to accept the participants' requests, such as not publishing their photographs. When collecting photographs by e-mail, it is necessary to consider how to confirm with the participants the use of their photographs.

#### 4.3.2.3 Spatially and Temporally Biased Data

When the site and period of photographs are not specified by ecologists, there is an advantage in that participants can participate at any time and place of their choice. Simultaneously, there is a disadvantage in that data with spatial and temporal biases are collected. Although the "*Hanamaru-Maruhana* national census" data are less biased than the GBIF data (Suzuki-Ohno et al. 2017), data bias must always be mitigated when using scientific research. Various methods have been proposed to reduce data bias (e.g., Syfert et al. 2013), and this problem can be solved technically.

### 4.3.3 *Lessons Learned from the* Hanamaru-Maruhana *National Census*

#### 4.3.3.1 Collaboration with Other Organizations

Taking digital photographs is an easy method for volunteers to participate in community science programs. However, even in the case of photographs, it is difficult for elementary school students to participate in the program. Children must go to school from morning to evening on weekdays, and a few children have a hobby of cameras on holidays. In the "*Hanamaru-Maruhana* national census," children participated in this program when teachers provided guidance and lent digital cameras to children for school club activities. It is necessary to organize events in which teachers and students can participate as school activities or parents with children can participate during a vacation. If the aims of a community science program include environmental education for children, it is better to collaborate with educational institutions and museums at an early stage.

This program has multiple aims, some of which are difficult to achieve without certain collaborations. Ultimately, we aim to connect this program to develop conservation plans and enhance conservation activities for bumble bees. We should connect the conservation guidelines to more concrete actual activities with volunteers and other organizations.

#### 4.3.3.2 Each Volunteer Has Special Abilities

Participants in a community science program often include people who are conducting research or have past research experience: current students, researchers in companies, former researchers, and people who used to do research when they were students. Such individuals have a high ability to collect data for scientific research. Even if participants have no experience in research activities, hobbyists can sometimes surpass professional scientists. In this program, we found various participants with high ability: college students, researchers in companies, amateur researchers, and highly skilled photographers. Through communication between professional scientists and participants, it is possible to determine the capability of each participant and collaborate with them.

#### 4.3.3.3 Not Overoptimizing Increases Possibilities

Optimization is often helpful; however, overoptimization can limit what is possible within the context scientific research. We first wanted to investigate bumble bee distributions and estimate their range shifts. Achieving this aim requires a large amount of presence data in which the distance between adjacent data points is greater than 1 km. We discussed whether to ask the participants to collect data at distances greater than 1 km. We also discussed whether to set a limit on the number of photographs sent by participants who sent many photographs at the same location. We discussed them several times; however, we did not limit the data. As a result, a large number of photographs were collected, which are highly useful for deep learning.

## Appendix. Bumble Bees

### *The Species Composition of Bumble Bees in Japan*

The target species of the *Hanamaru-Maruhana* national census are bumble bees observed in Japan. Sixteen bumble bee species including one kleptoparasitic species and one alien species inhabit the Japan archipelago excluding the Kurile Islands (Fig. 4.1, Tables 4.1 and 4.2, https://hanamaruproject.s1009.xrea.com/hanamaru_project/identification_E.html). Most of them also inhabit East or North Asia, and seven of these species also inhabit Europe. *Bombus honshuensis* and *B. yezoensis* are the endemic bumble bee species in Japan; however, *B. yezoensis* was sometimes treated as the synonym of *B. consobrinus* (e.g., Williams 1998). From a global perspective, the number of bumble bee species inhabiting the Japan archipelago excluding the Kurile Islands is relatively small. The number of bumble bee species is reported to be 66–70 in Europe and 50 in North America (Williams and Jepsen 2021).

The major species inhabiting the main island (Honshu) are *B. ardens*, *B. diversus*, *B. ignitus*, *B. hypocrita*, *B. honshuensis*, and *B. beaticola* (Fig. 4.1). *Bombus ardens* and *B. ignitus* are often observed at low elevations and *B. diversus* at middle-elevations. *Bombus hypocrita* and *B. honshuensis* at middle and high elevations and *B. beaticola* at high elevations. Different subspecies are observed in the northern

**Table 4.2** Bumble bee species and subspecies in the Japanese archipelago excluding the Kurile Islands

| Genus | Species | Subspecies | Author of scientific name | Synonym |
|---|---|---|---|---|
| *Bombus* | *cryptarum* | | Fabricius | *florilegus* |
| | *hypocrita* | *hypocrita* | Pérez | |
| | *hypocrita* | *sapporoensis* | Cockerell | |
| | *ignitus* | | Smith | |
| | *terrestris* | | Linnaeus | |
| | *consobrinus* | *wittenburgi* | Vogt | |
| | *diversus* | *diversus* | Smith | |
| | *diversus* | *tersatus* | Smith | |
| | *ussurensis* | | Radoszkowski | |
| | *yezoensis* | | Matsumura | *consobrinus* |
| | *norvegicus* | *japonicus* | Yasumatsu | |
| | *ardens* | *ardens* | Smith | |
| | *ardens* | *sakagamii* | Tkalcu | |
| | *ardens* | *tsushimanus* | Sakagami et Ishikawa | |
| | *beaticola* | *beaticola* | Tkalcu | |
| | *beaticola* | *moshkarareppus* | Sakagami et Ishikawa | |
| | *hypnorum* | *koropokkrus* | Sakagami et Ishikawa | |
| | *deuteronymus* | *deuteronymus* | Schulz | |
| | *deuteronymus* | *maruhanabachi* | Sakagami et Ishikawa | |
| | *honshuensis* | *honshuensis* | Tkalcu | |
| | *honshuensis* | *tkalcui* | Sakagami et Ishikawa | |
| | *pseudobaicalensis* | | Vogt | |
| | *schrencki* | *albidopleuralis* | Skorikov | |

island (Hokkaido) and the western island (Tsushima) of the Japanese archipelago. *Bombus ardens sakagamii*, *B. diversus tersatus*, *B. hypocrita sapporoensis*, *B. honshuensis tkalcui*, *B. beaticola moshkarareppus* inhabit only Hokkaido, whereas *B. ardens tsushimanus* inhabits only Tsushima.

According to the IUCN Red List criteria (implemented by the insect specialist group organized by the Ministry of Environment, Japan), *B. ignitus* and *B. cryptarum* are near-threatened (NT), and *B. consobrinus wittenburgi* and *B. ussurensis* are data-deficient (DD) (Fig. 4.1). *Bombus ignitus* is one of the major species in the main island (Honshu), however, defined as NT because of the risk that the populations of *B. ignitus* are declining in some areas. *Bombus cryptarum* inhabits only the eastern area of Hokkaido, whereas *B. consobrinus wittenburgi* and *B. ussurensis* inhabit only the central region of Honshu. Although not listed on the IUCN Red List, *B. deuteronymus maruhanabachi* and *B. norvegicus japonicus* inhabit only the central region of Honshu, including *B. consobrinus wittenburgi* and *B. ussurensis*. On the Red List defined by Nagano Prefecture, which is located in the central region of Honshu, *B. deuteronymus maruhanabachi* and *B. ussurensis* are vulnerable (VU), *B. ignitus* is NT, and *B. consobrinus wittenburgi* and *B. norvegicus japonicus* are

DD. Based on these criteria, we defined that rare species/subspecies are *B. ignitus*, *B. cryptarum*, *B. consobrinus wittenburgi*, *B. ussurensis*, *B. deuteronymus maruhanabachi*, and *B. norvegicus japonicus* in Suzuki-Ohno et al. (2021a).

## The Color Patterns of Bumble Bees

Bees, wasps, and hoverflies are difficult for nonexperts to identify species because they involve species with interspecific similarities (Fig. 4.10). One of the major causes of interspecific similarity is Batesian mimicry (Bates 1862). Hoverflies are often similar to bees and wasps, and some hoverflies with black and yellow hairs (e.g., *Merodon equestris*) are especially similar to bumble bees. Mimic hoverflies can be distinguished from bees by their face and body proportion. Another major cause of interspecific similarity is Müllerian mimicry (Müller 1878). The black and yellow patterns of bees and wasps are considered as Müllerian mimicry. Carpenter bees such as *Xylocopa appendiculata circumvolans* (lower right in Fig. 4.10) are often misidentified as bumble bees by volunteers. Mimicry within the same genus makes it quite difficult to distinguish between species. A black body with an orange tail like *B. ardens ardens* and *B. ignitus* (Fig. 4.1b and d) was considered as Müllerian mimicry within the same genus (Plowright and Owen 1980; Williams 2007; Hines

**Fig. 4.10** Species that volunteers often misidentified as bumble bees. Upper left: hoverfly (Photographer: Y. Suzuki-Ohno), upper right: *Eucera* sp. (Photographer: H. Morishima), lower left: *Apis mellifera* (Photographer: H. Morishima), and lower right: *Xylocopa appendiculata* (Photographer: H. Morishima)

**Fig. 4.11** Color patterns of male bumble bees (after Suzuki-Ohno et al. 2018). *Bombus ardens*, *B. hypocrita*, and *B. ignitus* show clear sexual dimorphism. *Bombus beaticola*, *B. hypnorum* and *B. ussurensis* show sexual dimorphism that male's hair is slightly yellowish (male of *B. ussurensis* is not shown in this figure). Upper left: male of *B. ardens* (Photographer: H. Morishima), upper right: male of *B. hypocrita* (Photographer: H. Morishima), lower left: male of *B. ignitus* (Photographer: Myano), and lower right: males of *B. beaticola* and *B. hypnorum* (Photographer: Y. Ishimori)

and Williams 2012), which is difficult to distinguish even by biologists (Suzuki-Ohno et al. 2022).

In addition, the intraspecific variation in bumble bees is large among bees. There are three major causes for intraspecific variation: castes, phenotypic variation, and subspecies. Bumble bees have the queen, worker, and male castes, and there are obvious differences between castes. The differences in castes are frequently remarkable in body sizes and sometimes in hair color. The hair colors of *B. ardens ardens*, *B. hypocrita*, and *B. ignitus* show clear sexual dimorphism (Figs. 4.1 and 4.11). Even in the same caste (especially worker), body size and color differ among individuals based on growing situations and age (Goulson 2009). For example, the hair color of *B. diversus* becomes yellowish when they inhabit colder regions, and their hair color becomes dull as they age. Two subspecies of *B. ardens* (*B. ardens sapporoensis* and *B. ardens tsushimanus*) have different hair colors from the nominotypical subspecies *B. ardens ardens*. Female *B. ardens sapporoensis* has dark yellow band on upper thorax and cream band on the abdomen, and female *B. ardens tsushimanus* has creamy-yellow body.

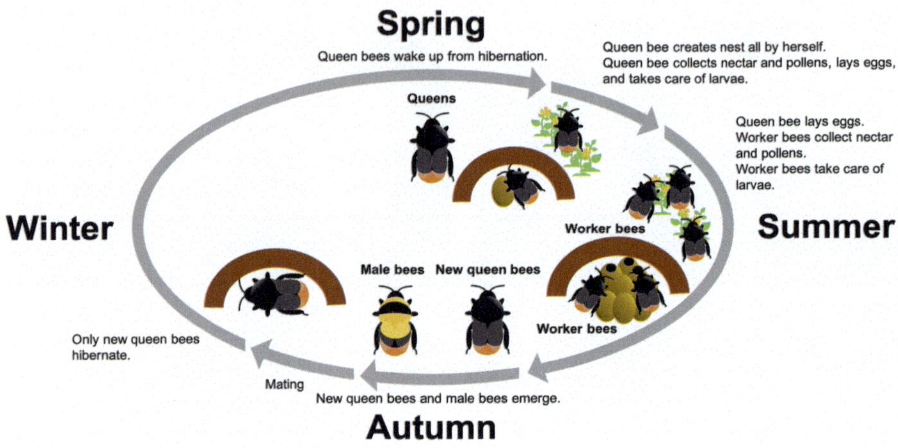


**Fig. 4.12** Life history of bumble bees

## *The Life History, Nesting, and Foraging Sites of Bumble Bees*

In Japan, the active season of bumble bees ranges from early spring to autumn. The active season of *B. ardens* starts in early spring and ends in early summer; a queen emerges in March, and the colony collapses in July. The active season of the other bumble bee species usually starts in spring and ends in autumn. In spring, a queen wakes up from hibernation, builds their nest, collects nectar and pollen, lays eggs, and raises larvae that become workers (Fig. 4.12). Once the larvae become adult workers, they collect nectar and pollen and nurture larvae, while the queen concentrates on laying eggs. Later in the active season, the colony produces new queens and males. Only the newly mated queen survives and hibernates until next year's active season.

Nesting sites of bumble bees are usually difficult to find. The locations of nesting sites often change because bumble bee colonies collapse within a year. Nesting sites are old rat nests underground, between rocks, and in tree hollows in forests. *Bombus deuteronymus* and *B. pseudobaicalensis* build aboveground nests covered with dry grass in the grasslands. Although they nest in natural and seminatural areas, *B. ardens* is adapted to urban areas and often builds nests under the floors of houses, birdhouses, and nearby graves. *Bombus hypnorum koropokkrus* also sometimes builds nests in artificial structures.

Foraging sites of bumble bees are often easy to find. Their foraging sites include flowers in parks, gardens, roadside trees, and crop fields, in addition to places with abundant nature. *Bombus ignitus* is sometimes observed along the coast. *Bombus hypocrita*, *B. honshuensis*, and *B. beaticola* are observed on mountain trails, highland flower fields, and ski resorts at high elevations in Honshu.

## References

Bates HW (1862) Contributions to an insect fauna of the Amazon valley. Lepidoptera: Heliconidae. Trans Linn Soc Lond 23:495–566

Cameron SA et al (2011) Patterns of widespread decline in north American bumble bees. Proc Natl Acad Sci U S A 108:662–667. https://doi.org/10.1073/pnas.1014743108

Chollet F (2017) Xception: deep learning with depthwise separable convolutions. In: Proc IEEE Conf Comput Vis Pattern Recognit, pp 1251–1258

Goulson D (2009) Bumblebees: behaviour, ecology, and conservation. Oxford University Press, New York

Hines HM, Williams PH (2012) Mimetic colour pattern evolution in the highly polymorphic *Bombus trifasciatus* (Hymenoptera: Apidae) species complex and its comimics. Zoological Journal of the Linnean Society 166(4):805–826. https://doi.org/10.1111/j.1096-3642.2012.00861.x

Inoue MN, Yokoyama J (2010) Competition for flower resources and nest sites between *Bombus terrestris* (L.) and Japanese native bumblebees. Appl Entomol Zool 45:29–35

Inoue MN, Yokoyama J, Washitani I (2008) Displacement of Japanese native bumblebees by the recently introduced *Bombus terrestris* (L.) (Hymenoptera: Apidae). J Insect Conserv 12:135–146

Jacobson MM, Tucker EM, Mathiasson ME, Rehan SM (2018) Decline of bumble bees in north-eastern North America, with special focus on *Bombus terricola*. Biol Conserv 217:437–445

Kerr JT et al (2015) Climate change impacts on bumblebees converge across continents. Science 349:177–180

Ministry of the Environment, and Ministry of Agriculture, Forestry and Fisheries (2017) Policy for use of alternative species of *Bombus terrestris*. (in Japanese) https://www.env.go.jp/nature/intro/2outline/attention/file/seiyouhoushin.pdf

Müller F (1878) Über die vortheile der mimicry bei schmetterlingen. Zool Anz 1:54–55

Nieto A et al (2014) European red list of bees. Publication Office of the European Union, Luxembourg

Oquab M, Bottou L, Laptev I, Sivic J (2014) Learning and transferring mid-level image representations using convolutional neural networks. In: Proc IEEE Conf Comput Vis Pattern Recognit, pp 1717–1724

Osborne JL et al (1999) A landscape-scale study of bumble bee foraging range and constancy, using harmonic radar. J Appl Ecol 36:519–533

Perez L, Wang J (2017) The effectiveness of data augmentation in image classification using deep learning. arXiv. 1712.04621

Phillips SJ, Anderson RP, Schapire RE (2006) Maximum entropy modeling of species geographic distributions. Ecol Model 190:231–259

Plowright RC, Owen RE (1980) The evolutionary significance of bumble bee color patterns: a mimetic interpretation. Evolution 34:622–637

Rasmont P et al (2015) Climatic risk and distribution atlas of European bumblebees. Biorisk 10:1–246

Soroye P, Newbold T, Kerr J (2020) Climate change contributes to widespread declines among bumble bees across continents. Science 367:685–688

Steffan-Dewenter I, Münzenberg U, Bürger C, Thies C, Tscharntke T (2002) Scale-dependent effects of landscape context on three pollinator guilds. Ecology 83:1421–1432

Suzuki-Ohno Y, Yokoyama J, Nakashizuka T, Kawata M (2017) Utilization of photographs taken by citizens for estimating bumblebee distributions. Sci Rep 7:11215

Suzuki-Ohno Y, Yokoyama J, Nakashizuka T, Kawata M (2018) Collection of organism observation data from photographs taken by citizens, problems and solutions. (in Japanese). e-Shuseibutsugaku Kenkyu, 2. https://www.speciesbiology.org/docs/e-book2_Ohno_etal_20181016.pdf. Accessed 11 Jan 2023

Suzuki-Ohno Y, Yokoyama J, Nakashizuka T, Kawata M (2020a) Estimating possible bumblebee range shifts in response to climate and land cover changes. Sci Rep 10:19622

Suzuki-Ohno Y, Kawaguchi L, Kobayashi C (2020b) Wild bee pamphlet. (in Japanese). https://sites.google.com/site/pollinationservicessociety/home/guideline. Accessed 11 Jan 2023

Suzuki-Ohno Y, Yokoyama J, Nakashizuka T, Kawata M (2021a) Bee occurrence data collected in citizen science program "*Hanamaru-Maruhana* national census" in Japan. Ecol Res 36:1028. https://doi.org/10.1111/1440-1703.12261

Suzuki-Ohno Y, Mitsuhata M, Nakamura J (2021b) Guidelines for giving habitat to bees. (in Japanese). https://sites.google.com/site/pollinationservicessociety/home/guideline. Accessed 11 Jan 2023

Suzuki-Ohno Y, Yokoyama J, Nakashizuka T, Kawata M (2022) Deep learning increases the availability of organism photographs taken by citizens in citizen science programs. Sci Rep 12:1210

Syfert MM, Smith MJ, Coomes DA (2013) The effects of sampling bias and model complexity on the predictive performance of MaxEnt species distribution models. PLoS One 8:e55158. https://doi.org/10.1371/journal.pone.0055158

Williams P (1998) An annotated checklist of bumble bees with an analysis of patterns of description (Hymenoptera: Apidae, Bombini). Bulletin of the Natural History Museum (Entomology) 67:79–152. The checklist is also available at https://www.nhm.ac.uk/research-curation/research/projects/bombus/subgenericlist.html

Williams P (2007) The distribution of bumblebee colour patterns worldwide: possible significance for thermoregulation, crypsis, and warning mimicry. Biol J Linn Soc 92:97–118

Williams PH, Jepsen S (eds) (2021) Bumblebee Specialist Group Report 2020

Williams PH, Osborne JL (2009) Bumblebee vulnerability and conservation world-wide. Apidologie 40:367–387

Yosinski J, Clune J, Bengio Y, Lipson H (2014) How transferable are features in deep neural networks? Adv Neural Inf Proces Syst 27:3320–3328

# Chapter 5
# Cyberforest Using Video and Sounds

**Kazuhiko W. Nakamura**

**Abstract** This chapter introduces the audio census, a bird monitoring by the non-profit organization Bird Research, using audio streaming provided by the Cyberforest project. During the first 3 years of the audio census that started in 2011, a comparison with the traditional field census confirmed that the audio census in a remote location can replace a large portion of the field census, which made the audio census continue uninterruptedly for 13 years. Multiple factors have contributed to the program's continuity for more than a decade. The aims of the Cyberforest are not tied solely to community science, which paradoxically gives it continuity as a community science program. One of the methodologies for community science to gain continuity may be the creation of a platform, where a variety of stakeholders, including those outside of community science, can participate.

**Keywords** Bird monitoring · Audio streaming · Cyberforest

## 5.1 What Is the Cyberforest?

### *5.1.1 Aims and Procedure*

Cyberforest is a conceptual framework that is a virtual forest on the Internet where people in remote areas can obtain local information as well as information in the field. It was proposed by Akio Fujiwara, one of the Cyberforest research project members. To achieve this concept, aesthetic information that can be captured by the five senses is indispensable. Aesthetic information here is sensory information that cannot be articulated in language, which is recorded in a way that people can intuitively grasp, such as images and sounds.

K. W. Nakamura (✉)
Graduate School of Frontier Sciences, The University of Tokyo, Tokyo, Japan
e-mail: k_nakamura@edu.k.u-tokyo.ac.jp



Y. Suzuki-Ohno (ed.), *Community Science in Ecology*, Ecological Research Monographs, https://doi.org/10.1007/978-981-97-0304-3_5

The Cyberforest started in 1997 as an information communication technology-based research project with the following primary aims: (1) to capture sound and imagery from forests in remote areas of Japan on a daily basis via unmanned cameras and microphones over an extended period of time and (2) to share the data obtained over the Worldwide Web (Saito et al. 2015). Although these are far from fully achieving the concept of aesthetic information of remote fields being available, maintenance of related equipment for long-term and continuous operation has been undertaken down to the present as the first process of the Cyberforest to date.

We, the Cyberforest Research Project, first set out to distribute and record images and sounds at the University of Tokyo Chichibu Forest (Fig. 5.1). The Chichibu Forest is uninhabited and has no access to a commercial power supply. It is the essence of the Cyberforest that such a truly "natural" forest is targeted.

First, we began recording images ahead of time in October 1995. Fujiwara, who is the proponent of the Cyberforest concept, installed a gasoline-powered generator next to the camera and went to the site every day to turn the generator to take video images. Obviously, it was difficult to continue this procedure over a long period, so he modified the system to automatically start and stop the generator at regular intervals each day and began filming unmanned. Considering the limitations of fuel and recording tape capacity, he decided to record only approximately 30 min a day, starting at approximately 11:30 a.m., and to visit the site every other week to refuel and change tapes.

The next step in 1998 was to start recording, and the noise of the generator engine became a problem. The solution was simply to relocate the generator to a location approximately 200 m away from the filming location, where the sound would not reach. This made it possible to record the sounds of nature. However,

**Fig. 5.1** Fixed-point camera and microphone at the University of Tokyo Chichibu Forest, first installed by the Cyberforest project

since the operational status of the system could not be monitored except on the occasion of biweekly visits to the site, minor errors in maintenance work often caused the recording to be interrupted for approximately 2 weeks. This was another reason why we needed to establish an Internet connection at filming locations.

It was in October 2007, 12 years after we started filming, that we installed an Internet connection at the filming location in the Chichibu Forest. Using a cell phone line with an uplink speed of 64 Kbps, low-quality video with sound at a resolution of QVGA (320 × 240 pixels) for only 10 s was delivered on the same day to a server at the University of Tokyo. Despite the low quality of the video, we could observe the trees changing their leaf colors day by day, even from a remote location such as the University of Tokyo campus. When the Internet connection was installed, we also improved the power supply by installing solar panels.

In 2010, we established a parabolic antenna for a satellite Internet connection at the location, and a 2 Mbps uplink broadband connection became available. Using this line, the company introduced high-resolution photography using a digital SLR camera and began streaming higher-quality audio. Thus, images are immediately transmitted as files, and sound is recorded on the university's server side, which eliminates the need for on-site recording media and realizes continuous, unmanned aesthetic information recording in areas with no power supply.

We have thus accumulated various kinds of know-how in the Chichibu Forest for more than 20 years and began to expand the number of monitoring sites around 2011, taking advantage of this know-how. As of March 2023, we are also distributing and recording images and sounds at the University of Tokyo Hokkaido Forest (Furano City), the University of Tokyo Fuji Iyashinomori Woodland (Yamanakako Village), and Shinshu University Shiga Nature Education Park (Yamanouchi Town).

### *5.1.2 Comprehensive Applications*

The aims of the Cyberforest are not limited to community science but are a comprehensive practice that explores knowledge to create a new contemporary relationship between human society and the natural environment. This chapter presents an audio census of birds as an example of an application that can be captured in the context of community science, but before describing it in detail, other applications are also briefly noted below.

First, the recording of video and sound allows for temporal observation of forest landscape change. In particular, tree phenology can be determined to a certain accuracy even with low-quality video. For example, the full bloom date in *Prunus verecunda* and the bud flush date in *Fagus japonica* were able to be determined more unambiguously than by visiting the site from remote locations and visually observing them (Nakamura et al. 2011). Regarding this application, in principle, it is possible to quantify tree phenology through mechanical image analysis, but for the Cyberforest, the quantification should address the relationship between human society and the natural environment. With this in mind, we have been mainly engaged in application practices that focus on the significance of people observing tree phenology with their eyes through visual images rather than mechanical quantification.

The process in which people look at their relationship with the natural environment through the observation of video and sound can be captured in the context of environmental learning. The first application practice related to environmental learning was the production of multimedia content of forest landscape changes over the passage of seasons organized in paper form and on DVDs (Saito et al. 2006). Subsequently, based on these contents, we developed learning materials through discussions with school teachers and accumulated practical cases in which these materials were actually used in classes at schools (Nakamura et al. 2019). There have been cases of utilization in a wide range of age groups, from elementary school

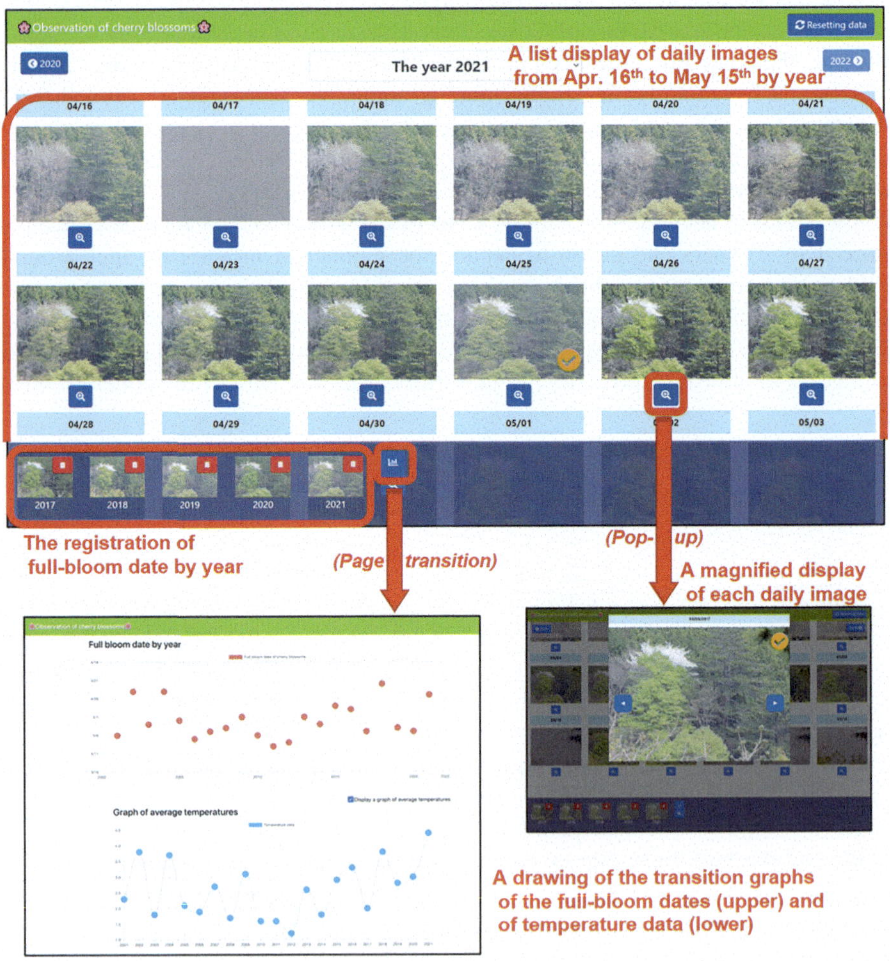


**Fig. 5.2** On-screen display of the web application for retrospective observation of cherry blossom phenology using fixed-point images by the Cyberforest

students to adults. Among these educational materials, a web application for retrospective observation of cherry blossom phenology (Fig. 5.2) is available on the Internet (https://pheno.cf4ee.jp/yatake_sakura/index_en.html) and can be used freely by anyone.

In addition, sounds suggest the possibility of broader applications. A private company, inspired by Cyberforest's 2010 startup of live audio streaming, launched a similar service in March 2013 with our cooperation and has continued the service to the present day in 2023. This service is not dedicated only to the context of environmental learning but also embraces a variety of usages, such as healing, entertainment, and artistic appreciation. Meanwhile, sound designer Haruo Okada has been extracting characteristic sounds from Cyberforest's vast sound archive that has been accumulated since 1998 and has started to distribute them through a music subscription service in February 2020. Moreover, an experimental attempt to capture the effects of listening to natural sounds on human psychological and physiological responses was initiated by an affiliated group at the Cyberforest (Suko et al. 2022).

As described above, the Cyberforest has a ripple effect on a wide range of fields, including education, arts, and healthcare, and various applications are being promoted through collaboration between researchers and practitioners. The Cyberforest has held its annual symposium continuously since 2012, and through this opportunity, continuous collaboration among various stakeholders has been realized. As described in detail in the next section, the audio census is one such multistakeholder collaborative project.

### *5.1.3 Target Species, Recording Method, and Types of Volunteers on the Audio Census*

The Audio Census is a program led by Bird Research, a Japanese nonprofit organization, to monitor forest birds using live audio streaming. Bird Research is an organization that promotes community science, with one of its goals being to train amateur bird researchers. As part of its activities, it had previously installed IC recorders in forests around Japan to monitor the chirping season of forestial birds, which are strongly affected by climate change. As a derivative activity, they began a real-time audio census in 2011, using the high-quality audio streaming that the Cyberforest has been conducting since 2010.

The audio census of birds is conducted from April 1 to June 30 each year, from 10 min before to 1 h after sunrise on each day. The three target sites are listed below, and the census is conducted at one site per day in sequence, continuing the three-day cycle.

1. The University of Tokyo Chichibu Forest (Chichibu City)
2. The University of Tokyo Fuji Iyashinomori Woodland (Yamanakako Village) [and the University of Tokyo Hokkaido Forest (Furano City) for 15 min just before the census on Fuji Iyashinomori Woodland]

3. Shinshu University Shiga Nature Education Park (Yamanouchi Town)

Two all-weather splashproof microphones (Sony F-115B) are installed in the forest for stereo use (Fig. 5.3). The audio streaming of all of these locations is publicly available (on http://cyberforest.jp/), and anyone with an Internet connection is free to listen.

The census recording will be done using the Internet Relay Chat (IRC) framework. Program participants use an IRC client to connect to channel #BirdResearch for the audio census and record the species name and voice type heard every minute. All bird species are targeted for recording, mainly *Treron sieboldii*, *Poecile montanus*, *Periparus ater*, *Sittiparus varius*, *Parus minor*, *Sitta europaea*, *Larvivora cyane*, *Ficedula narcissina*, *Turdus chrysolaus*, *Zoothera aurea*, *Regulus regulus*, *Certhia familiaris*, *Zosterops japonicus*, *Turdus cardis*, *Horornis diphone*, *Phylloscopus coronatus*, *Phylloscopus borealoides*, *Phylloscopus xanthodryas*, *Chloris sinica*, *Streptopelia orientalis*, *Cuculus poliocephalus*, *Cuculus canorus*, *Cuculus saturatus*, *Tarsiger cyanurus*, *Emberiza variabilis*, *Hierococcyx hyperythrus*, *Urosphena squameiceps*, etc. For voice type, the codes S for songs, C for calls, and D for drumming are used. In addition to the bird voice records, small talk or question-and-answer conversations also take place in this chat.

There are no specific qualifications or skills required to participate in this voice census, and anyone with access to voice streaming and IRC can participate. Beginners can learn various bird sounds hands-on by simply listening to the audio streaming and browsing the bird voice records on IRC.

**Fig. 5.3** All-weather splashproof microphones (Sony F-115B) are used for the Cyberforest audio streaming at the University of Tokyo Hokkaido Forest

## 5.2 The Outcome of the Audio Census with Cyberforest

During the first 3 years of the audio census of birds, which began in 2011, the main consideration was a comparison with the traditional field census conducted in person at the survey sites. According to Saito et al. (2015), the main results are as follows:

- The 20 most abundant species in the audio census accounted for 90% of the species recorded in the field census.
- Species with human-avoidance characteristics, such as *Syrmaticus soemmerringii*, are more likely to appear in the audio census.
- Species that are not particularly vocal, such as *Sittiparus varius* and *Parus minor*, are easier to detect with field censuses that provide visual cues.

As described above, despite minor differences, it was confirmed that the audio census in a remote location can replace a large portion of the field census.

Thus, due in part to the fact that the usefulness of the audio census was confirmed to a certain extent, the census continued uninterruptedly for 13 years from 2011 to 2023. This means that it is possible to capture the effects of climate conditions each year from specific aspects concerning a large number of bird species, such as those mentioned earlier. Since the sites of the audio census are all educational and research facilities of universities, where meteorological data and ecological data from researchers in various fields have been accumulated, it is expected that the data will be used not only for tracking trends over time but also for multifaceted and multilayered analysis.

In addition, the audio census produced outcomes beyond the field of ecology. The best example is the repeated participation in Reveil, a worldwide annual event concerning audio streaming. Reveil is an event where audio streaming enthusiasts gather at the Soundcamp venue in London to listen to audio streaming of sunrise in sequence around the world over a 24-h period and has been held annually in early May since 2014. While many participants are keenly interested in the ecological sounds heard on streaming from different locations, the main intent of the event is not ecological research but participatory artistic production. The audio census with Cyberforest streaming was first added to the Reveil program in 2017, and this continued every year thereafter through 2023, except for 2019. From 2020 onward, bird monitoring using IRC by members of the NPO Bird Research in the audio census was delivered to the Soundcamp venue on Reveil via Zoom webinar along with audio streaming (Fig. 5.4). This achievement demonstrates the value of the audio census as an artistic exhibition.

**Fig. 5.4** On-screen display of the audio census in Reveil 2021 delivered via Zoom Webinar

## 5.3 The Advantages, Disadvantages, and Lessons Learned from the Audio Census with Cyberforest

### *5.3.1 The Advantages of this Program*

If a situation can be created where power supply and network infrastructure are available, unmanned bird monitoring can be realized from remote locations. The capability to participate in surveys remotely via the Internet is important in the context of community science. This is because the skill of identifying the sounds of birds is not the same as the skill of safely walking through a rugged forest. The more difficult the site is to reach, the greater the advantage for the nonspecialist to focus only on identifying the sounds of birds. In addition, the capability to create a no-person situation has ecological significance in terms of capturing the true ecology of birds.

Furthermore, the capability to participate in a survey via the Internet makes it possible for multiple people to survey the same location at the same time. This not only enhances the accuracy of the survey through checks by multiple people but also has a positive effect on the motivation of the surveyors by allowing them to survey while engaging in textual chatter about the content of the survey on IRC, which ultimately enhances the continuity of the program. Actually, the chatting is very active, and I have confirmed that it has increased the motivation of the surveyors.

### 5.3.2 The Disadvantages of this Program

The capital investment required for live audio streaming of rugged forests, where the advantages described above are strongly exhibited, would require a relatively high cost. It may be too high a cost for the single aim of bird monitoring only. The Cyberforest has solved this problem in part by not limiting its purpose to bird monitoring alone, as mentioned above. It may be necessary to consider collaboration with other sectors to realize a program similar to the audio census.

While it would not be much of a problem in a rugged forest such as those targeted by Cyberforest, the installation of microphones in proximity to human societies would involve the risk of privacy. In this regard, there may be a way to use AI to automatically erase human voices, but this has not yet been put into practical use at this time. Under such circumstances, the audio census may be premised on limiting the target area to places where people are basically inaccessible.

Furthermore, it should be considered that the assumption on which the audio census by NPO Bird Research obtained results similar to the field census is that the main surveyors of the audio census had extensive experience with field censuses of the same locations. A skilled surveyors like them do not merely identify the sounds of birds in an audio census; they combine all the sounds they hear, including non-birds, with their own experiences to imagine the situation at the site. In training audio census surveyors, it is important to accumulate field experience in addition to the ability to identify the sounds of birds, which is not always easy.

### 5.3.3 Lessons Learned from the Audio Census

The description of the audio census with the Cyberforest thus far highlights that multiple factors have contributed to the program's continuity for more than a decade. On the surface, it seems that the program can be captured in the context of community science, but the inner workings may be slightly more complex.

For community science in general, continuity of monitoring is often an issue. The audio census is based on the Cyberforest, where the aims are not tied solely to community science, which paradoxically gives it continuity as a community science program. The main surveyors of the audio census have had the opportunity to interact with researchers and practitioners in other fields, such as education and arts, at the Cyberforest symposiums and other venues, which may be one of the motivating factors for the continuation of the audio census. One of the methodologies for community science to gain continuity may be the creation of a platform such as the Cyberforest, where a variety of stakeholders, including those outside of community science, can participate.

In addition, focusing on the value that community science programs themselves have beyond ecological monitoring may also lead to their continuity. As mentioned above, the audio census with the Cyberforest also has an aspect of participatory art.

The exhibition should not only be entertainment but also a glimpse into the profound world of academia. We may even find environmental learning value in presenting this to community science beginners as an informal learning experience.

## References

Nakamura K, Yoneya N, Saito K, Fujiwara A (2011) Unambiguous determination of full bloom date in Prunus verecunda and bud flush date in Fagus japonica based on video recording archives. Bull Tokyo Univ For 125:1–10

Nakamura KW, Fujiwara A, Kobayashi HH, Saito K (2019) Multi-timescale education program for temporal expansion in ecocentric education: using fixed-point time-lapse images for phenology observation. Educ Sci 9(3):190

Saito K, Fujiwara A, Toko A, Yano A, Okamoto T (2006) Design study of forest environment multimedia contents for environmental education with video data at the Tokyo University Forests in Chichibu. Bull Tokyo Univ For 116:267–281

Saito K, Nakamura K, Ueta M, Kurosawa R, Fujiwara A, Kobayashi HH, Nakayama M, Toko A, Nagahama K (2015) Utilizing the Cyberforest live sound system with social media to remotely conduct woodland bird censuses in Central Japan. Ambio 44:S572–S583

Suko Y, Shindo T, Saito K, Takayama N, Warisawa S, Sakuma T, Ito M et al (2022) Alleviating surgeons' stress through listening to natural sounds in a half-encapsulated rest space after an operation: a pilot, longitudinal field study. Int J Environ Res Public Health 19(19):12736

# Chapter 6
# Community Science Initiatives Utilizing Environmental DNA

**Michio Kondoh, Minoru Kasada, Takuzo Abe, Akihide Kasai, Akihiro Dazai, Reiji Masuda, Satoquo Seino, Shota Suzuki, Yukari Suzuki-Ohno, and Akifumi S. Tanabe**

**Abstract** The technology that utilizes environmental DNA (eDNA), which refers to DNA released by organisms into the environment, has rapidly advanced as a powerful tool for evaluating biodiversity within a targeted ecosystem. eDNA surveys involve minimal fieldwork, enabling collaboration with academia, nonprofit organizations (NPOs), and companies to conduct surveys at various locations and frequent intervals. In this chapter, we explore biodiversity monitoring using eDNA metabarcoding within the framework of community science. Initially, we introduce ANEMONE, an eDNA-based biodiversity monitoring network that has laid the

M. Kondoh (✉)
Graduate School of Life Sciences, Tohoku University, Sendai, Miyagi, Japan

Advanced Institute for Marine Ecosystem Change (WPI-AIMEC), Tohoku University, Sendai, Miyagi, Japan
e-mail: michio.kondo.b8@tohoku.ac.jp

M. Kasada
Graduate School of Life Sciences, Tohoku University, Sendai, Miyagi, Japan

Field Science Center for Northern Biosphere, Hokkaido University, Horokanai, Japan

T. Abe · S. Suzuki
Minamisanriku Nature Center, Minamisanriku, Miyagi, Japan

A. Kasai
Faculty of Fisheries Sciences, Hokkaido University, Hakodate, Hokkaido, Japan

A. Dazai
Center for Sustainable Society, Minamisanriku, Miyagi, Japan

R. Masuda
Maizuru Fisheries Research Station, Field Science Education and Research Center, Kyoto University, Kyoto, Japan

S. Seino
Graduate School of Engineering, Kyushu University, Fukuoka, Japan

Y. Suzuki-Ohno · A. S. Tanabe
Graduate School of Life Sciences, Tohoku University, Sendai, Miyagi, Japan



Y. Suzuki-Ohno (ed.), *Community Science in Ecology*, Ecological Research Monographs, https://doi.org/10.1007/978-981-97-0304-3_6

groundwork for eDNA-based community science in Japan. Then, we highlight several examples of cooperative eDNA surveys carried out with collaborators, including volunteers, governments, residents, shipping companies, and nature enthusiasts. We present the outcomes of implementing the ANEMONE framework. Finally, we delve into the potential future direction of biodiversity monitoring in conjunction with community science and its possible contributions to scientific advancement, industry, and addressing societal issues. We stress the crucial and significant role of community science-based eDNA monitoring in future ecosystem conservation through various means, as well as its potential to empower local communities by providing valuable ecosystem information.

**Keywords** Environmental DNA (eDNA) · eDNA metabarcoding · Biodiversity monitoring · Monitoring network · Database · Large-scale survey

## 6.1 Environmental DNA-Based Biological Surveys

Environmental DNA (eDNA) technology has rapidly developed to be a powerful tool for biological surveys that estimate the presence, distribution, and taxonomy of organisms and evaluate local biodiversity (Minamoto et al. 2016; Tsuji et al. 2019; Beng and Corlett 2020). eDNA refers to DNA that is shed from organisms into the environment, such as water, soil, and air, through various means, including feces, hair, mucus, gametes, and skin (Taberlet et al. 2012). An eDNA technology is a technological framework that enables estimating the ecosystem state from eDNA. Since the first report in 2008 demonstrating the detection of bullfrog (*Rana catesbeiana*) from its DNA in environmental water (Ficetola et al. 2008), eDNA technology has significantly developed and is now widely recognized as a cost-effective method for biological surveys.

Among various techniques in eDNA technology, eDNA metabarcoding is a promising method for assessing multiple species in the target ecosystem (Valentini et al. 2016; Deiner et al. 2017). The procedure of eDNA metabarcoding involves amplifying the DNA of multiple organism groups using PCR with primers designed for a DNA region shared by target groups (i.e., universal primer); the resulting amplicons are then sequenced by next-generation sequencing, and the obtained sequence information is compared to reference data to identify multiple organism groups present at the surveyed location. Following the first practice of eDNA metabarcoding detecting four fish species in rivers (Minamoto et al. 2012), this method rapidly expanded, and currently, eDNA metabarcoding techniques are applicable to various taxonomic groups, including fish (Miya et al. 2015; Yamamoto et al. 2017), mammals (Ushio et al. 2017), crustaceans (Komai et al. 2019), and birds (Ushio et al. 2018a), and have been proposed.

The process of eDNA metabarcoding in aquatic environments consists of multiple steps. A typical procedure of eDNA metabarcoding using active sampling is as follows.

1. Water sampling: Collecting water samples from the survey area. The required water volume varies depending on the habitat type, ranging from a few hundred milliliters to several liters.
2. Filtration: The actual form of eDNA in the environment is fragments of cells derived from organisms present in the water. These fragments, among others, are separated from the water samples using filters.
3. DNA extraction: Extracting DNA from the remaining cell fragments on the filters. This involves breaking down proteins and removing impurities present in the samples.
4. DNA amplification and library preparation: Amplifying eDNA to an analyzable quantity using polymerase chain reaction (PCR) and simultaneously adding various adapters to both ends of the PCR products to process them into molecules suitable for analysis by next-generation sequencers.
5. DNA sequencing: Determining the base sequences of PCR products obtained from eDNA using next-generation sequencers or similar technologies.
6. Bioinformatical data analysis: Identifying the types of organism groups by comparing the obtained base sequence information with DNA libraries.

Over the past decade, there has been notable progress in the field of eDNA metabarcoding, with a particular focus on fish detection. Among various techniques proposed, the MiFish method (Miya et al. 2015) has emerged as the leading approach, surpassing other methods in terms of its ability to amplify a diverse range of fish species (Collins et al. 2019; Zhang et al. 2020). This method has become widely accepted as the de facto standard in the field (Minamoto et al. 2021).

The MiFish approach employs universal primers designed to target a hypervariable region of the mitochondrial genome's 12S rRNA gene, specifically spanning from 163 to 185 base pairs. Miya et al. (2015) introduced this method and conducted empirical testing in four aquarium tanks at the Churaumi Aquarium. The results were successful in detecting species from 123 genera and 168 distinct species, which are the fishes actually kept in the tanks. Yamamoto et al. (2017) subsequently applied the MiFish method to 47 locations within a relatively enclosed bay, where they were able to detect a total of 128 fish species. Comparisons with previous scuba diving surveys conducted biweekly in the same bay since 2002 demonstrated that many previously identified species were also detected through eDNA surveys. Furthermore, some species, which had been observed only infrequently in the past, were included in the species detected through eDNA surveys. The accumulation of reference data for MiFish has also made significant progress. The MiFish database (MiFish DB Ver. 3.85) currently encompasses mitochondrial information for an impressive 9569 species, further solidifying the merits of this method (Zhu et al. 2023). Moreover, Ushio et al. (2018b) proposed the application of artificial standard DNA to MiFish-based eDNA metabarcoding, a development that allows for the quantification of eDNA. This innovation opens up the possibility of efficient quantitative eDNA-based monitoring of fish diversity.

In this chapter, our objective is to delve into the realm of biodiversity monitoring utilizing eDNA barcoding within the framework of community science. In Sect. 6.2, we will demonstrate the significance of eDNA monitoring in the natural sciences and its relevance in tackling societal challenges. We will also outline the pivotal role that community science can play in advancing these efforts. Sections 6.3 and 6.4 will shed light on specific cases, such as ANEMONE, the eDNA monitoring network implemented in Japan, and will highlight the valuable contributions from a diverse range of stakeholders, including local residents, government officials, and businesses. Lastly, in Sect. 6.5, we will explore the future prospects of biodiversity monitoring in tandem with community science, with a particular focus on the anticipated role it can play in addressing biodiversity-related issues.

## 6.2 Acquiring Large-Scale Ecosystem Information Through Community Science

Ecosystems are intricate, nonlinear-dynamical systems driven by the interactions among numerous species and environmental factors (Mittelbach and McGill 2019). The variations in species composition and interactions, contingent on specific contexts, give rise to a multitude of ecosystem functions. These, in turn, influence how ecosystems respond to disturbances and contribute to the ecosystem's stability (Ushio et al. 2018a, b, c; Catford et al. 2022). Within this intricate "biological machinery" that sustains human society, individual species, including humans, are potentially essential components that both constitute and rely on the ecosystem. A primary objective of ecology is to comprehend the structure, dynamics, functions, and relationships within these highly complex ecosystems, where each component and the entire system are intricately interconnected. To achieve this understanding, it is imperative to gather information on species composition, which illuminates when, where, and in what quantities organisms exist. Moreover, for the sustainable development of human society and economy, as well as the well-being of all individuals, it is crucial to harness ecosystems appropriately. However, this necessitates the accurate and swift assessment of ecosystem conditions and their changes. Data on species composition forms one of the foundational pieces of information about ecosystems, playing a pivotal role in both fundamental and applied ecology.

The primary advantage of eDNA surveys lies in their capacity to furnish extensive ecosystem information. Generally, comprehending the state, dynamics, responses, and future projections of intricate systems necessitates data that align with the complexity of the system in terms of both quality and quantity. Ecosystems represent intricate systems where numerous species interact within specific contexts, necessitating comprehensive observational data for a thorough understanding. eDNA surveys entail minimal fieldwork, allowing for surveys at multiple locations and frequent intervals. Furthermore, by employing eDNA metabarcoding, information pertaining to various biological communities spanning different taxonomic groups can be garnered. Another noteworthy advantage of eDNA surveys is their suitability for conducting extensive investigations. Ecosystems are frequently influenced by processes that occur on a large spatial scale. For instance, migratory birds and fishes are known to cover substantial distances, often spanning thousands of kilometers (Alerstam et al. 2003). This implies that the presence and dynamics of these organisms in one location may be impacted by processes occurring in distant places, thousands of kilometers away (Bauer and Hoye 2014; Moisan et al. 2023; Singh et al. 2021). With proper implementation, eDNA holds the potential to deliver comprehensive insights into ecosystems on a grand scale, employing a unified methodology.

In the subsequent sections, we will delve into the field of community science employing eDNA in Japan and examine its significance. To begin, we will introduce ANEMONE, the eDNA-based biodiversity monitoring network, which has laid the groundwork for eDNA-based community science in Japan. Following that, we will spotlight the cooperative eDNA surveys carried out in conjunction with local communities and the results stemming from the implementation of the ANEMONE framework. Lastly, we will explore the prospective directions of this program and the contributions it can make to scientific progress, industry, and the resolution of societal issues.

## 6.3 ANEMONE: eDNA Monitoring Network

In Japan, the first, large-scale community science utilizing eDNA was launched in 2021 thanks to a collaboration involving academia, nonprofit organizations (NPOs), and companies. This milestone was achieved through the establishment of the All Nippon eDNA Monitoring Network, more commonly referred to as ANEMONE. ANEMONE is the product of collaborative efforts between industry, government, and academia. In the following sections, we will elucidate the inner workings and objectives of ANEMONE.

The predecessor of ANEMONE was an eDNA survey conducted as part of a large-scale project (JST CREST: "Development of Quantitative Monitoring and Ecosystem Assessment Methods for Fish Community Based on Environmental DNA Analysis") targeting the coastal areas nationwide in Japan. Building upon the established survey methods during that time, the Environmental DNA Monitoring Network, ANEMONE, was established in 2019. ANEMONE initially started through the collaboration between universities and research institutions in response to a call from Tohoku University. Tsukuba University plays a central role, coordinating observation points, preparing and distributing survey kits, and receiving eDNA samples. Kazusa DNA Research Institute conducts DNA analysis, while Tohoku University oversees bioinformatical data analysis and overall coordination. Subsequently, researchers in Japan Fisheries Research and Education Agency (FRA) joined the project, contributing to field sampling and conducting certain wet analyses. ANEMONE initially focused on fish as the primary target and implemented eDNA metabarcoding using MiFish. More than 1900 surveys conducted as of October 7, 2023, resulting in the detection of 1086 taxonomical species (of which

682, 295, and 98 are identified to species, genus, and family, respectively), which consist of 682 species, 569 genera, and 202 families in total. Extracted DNA samples are stored for future reanalysis. Furthermore, the acquired data is managed in a dedicated database, as will be described in detail later.

ANEMONE is characterized by its inclusivity, welcoming participation from various entities, and its data openness.

Inclusive observation: ANEMONE has implemented a strategic division of labor to facilitate the engagement of diverse institutions and individuals without necessarily requiring specialized knowledge or skills. Specifically, the following three tasks are carried out by different entities: (a) water collection and filtration in the field, (b) wet experimental work from DNA extraction to sequencing, and (c) bioinformatics analysis and data management. Among these tasks, (a) field-based water collection and filtration are currently accessible to nonprofessional scientists and are primarily conducted by three distinct types of entities. The first category involves regular observations conducted at fixed monitoring sites across the country, ranging from weekly to seasonal surveys. Participants in these observations include universities, national research institutions, and representatives from local governments. The second category consists of one-time volunteer surveys, although some volunteer groups opt to repeat these surveys at the same locations, as noted by Suzuki-Ohno et al. (2023). The majority of participants in these surveys hail from local communities, including citizens and nonprofit organizations (NPOs). The third category comprises paid surveys conducted by companies and other organizations. Entities interested in conducting eDNA surveys using the ANEMONE approach can participate and acquire data by covering the associated expenses.

Open data: All collected data is stored in the dedicated ANEMONE DB (https://db.anemone.bio), a specialized database. The operation of ANEMONE DB is guided by an open data policy, allowing unrestricted access for individuals and organizations to utilize the data for both academic and commercial purposes. The database encompasses sequence data derived from detected eDNA, occurrence data of taxonomical species inferred from these sequences, as well as valuable metadata including experimental parameters, sampling locations, and time of collection. By default, all survey data acquired through ANEMONE is made publicly available. Nevertheless, in order to encourage scientists' involvement, a 6-month period of data exclusivity is granted to personnel responsible for water sampling. During this time frame, they have the opportunity to use the data for the preparation of scientific papers or for presenting their findings at conferences and other relevant events.

## 6.4 Contributions of the Local Community to the ANEMONE Project

### *6.4.1 ANEMONE Observation by Community Volunteers*

The involvement of the local community in contributing to the ANEMONE biodiversity survey commenced in 2020 through a program, "Fish Survey Project Using Environmental DNA," which was operated as a collaboration between ANEMONE and the nonprofit organization Earthwatch Japan. Earthwatch Japan is an NPO dedicated to supporting community science, offering citizens the chance to partake in scientific activities while also providing scientists with volunteer assistance and the means to disseminate their research to the public. The community participation program within ANEMONE sets forth three primary objectives for citizen volunteers. Firstly, it aims to comprehensively document the biodiversity of fish along the Japanese coast, achieving an unprecedented level of resolution in this endeavor. Secondly, it aspires to establish a global biodiversity database that is accessible to scientists worldwide, promoting open access to valuable information. Lastly, the program provides participants with the opportunity to enhance their understanding of the state of their local ecosystems and lays the groundwork for contemplating strategies to conserve and sustainably utilize Japan's coastal areas.

In this program, akin to various other eDNA citizen surveys (Biggs et al. 2015; Larson et al. 2017; Tøttrup et al. 2021), volunteers actively contribute to ANEMONE by engaging in water sampling and filtration processes (Fig. 6.1). As mentioned earlier, ANEMONE streamlines its surveys by distributing different responsibilities among various institutions, namely, water sampling and filtration, wet analysis, and dry analysis. The survey methodology involves community volunteers taking on the

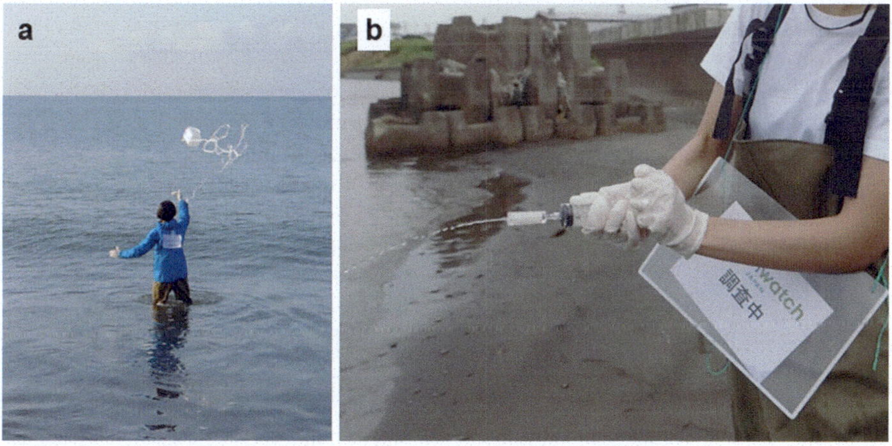


**Fig. 6.1** Water sampling (**a**) and filtering (**b**) conducted by volunteers in the program, "Fish Survey Project Using Environmental DNA"

tasks of water sampling and filtration, which were previously conducted by university marine stations and other entities during routine observations at fixed monitoring points. These coastal ecosystem survey locations are typically selected by volunteers based on their individual interests. By allowing participants to pursue surveys driven by their own curiosity, it is anticipated that their motivation will be significantly enhanced.

In this program, local community participants conduct surveys based on provided sampling kits, manuals, and video guides (Fish Survey Project Using Environmental DNA 2020; see Fig. 6.2) without the need for direct assistance from researchers. Although professional scientists organize this survey, they provide detailed explanations of the survey methods in advance during guidance sessions, without offering on-site support to citizen volunteers. This approach was implemented due to the impact of the COVID-19 pandemic that began in 2019. Traditionally, Earthwatch's activities involved citizens visiting researchers' study sites and assisting in surveys, directly guided by researchers. However, the constraints imposed by the COVID-19 pandemic made it challenging for scientists and volunteers to convene at the same location and collaborate in person. Consequently, a program for citizen participation in eDNA surveys was initiated, enabling citizen participants to contribute to scientists' ecological survey activities in smaller, socially distanced groups.

Conducting surveys relying solely on manuals without the direct involvement of researchers in locations chosen by citizens themselves presents a challenging endeavor. The primary concern in this context is the risk of contamination. In eDNA surveys, the DNA of fish is amplified and analyzed from exceedingly minute quantities present in the water. Consequently, if human-derived DNA contaminates the samples, it can introduce unwanted observational noise into the data. To mitigate this risk, various precautions are rigorously implemented in eDNA surveys to prevent contamination. For instance, all equipment is meticulously cleaned with bleach during the survey, and individuals collecting water samples are required to wear gloves. Moreover, separate rooms are designated for DNA extraction and DNA amplification to ensure strict segregation and minimize the possibility of contamination. The most up-to-date and stringent measures are consistently enforced. Maintaining a vigilant focus on preventing contamination is regarded as the second most critical safety measure in this community survey, following safety management.

The survey findings are disseminated through online reporting sessions attended by both volunteer participants and professional scientists. These sessions encompass comprehensive discussions on various topics, including the observed patterns in fish species composition across the nation and their ecological significance. They also provide valuable opportunities for constructive dialogues between citizen volunteers and professional scientists. Over the span of 3 years, from 2020 to 2022, the program witnessed the participation of 341 volunteers, and a total of 383 surveys were conducted (Fig. 6.3). Remarkably, the surveys yielded the detection of 797 taxonomical species (Fig. 6.4), including rare and endangered ones such as Japanese eels (*Anguilla japonica*) and Japanese lates (*Lates japonicus*). All of this invaluable data is made publicly accessible through the dedicated ANEMONE database,

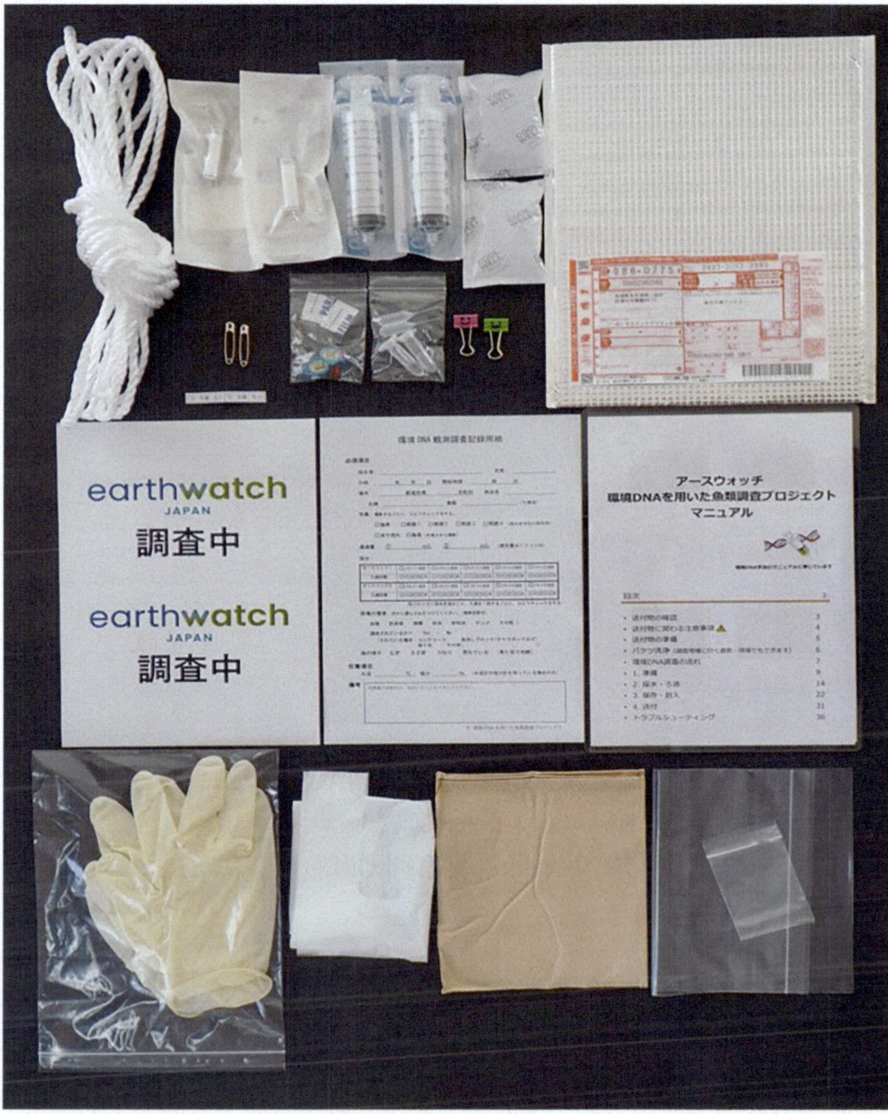


**Fig. 6.2** The eDNA sampling kits distributed to volunteer participants in the community science-based eDNA survey program (Suzuki-Ohno et al. 2023)

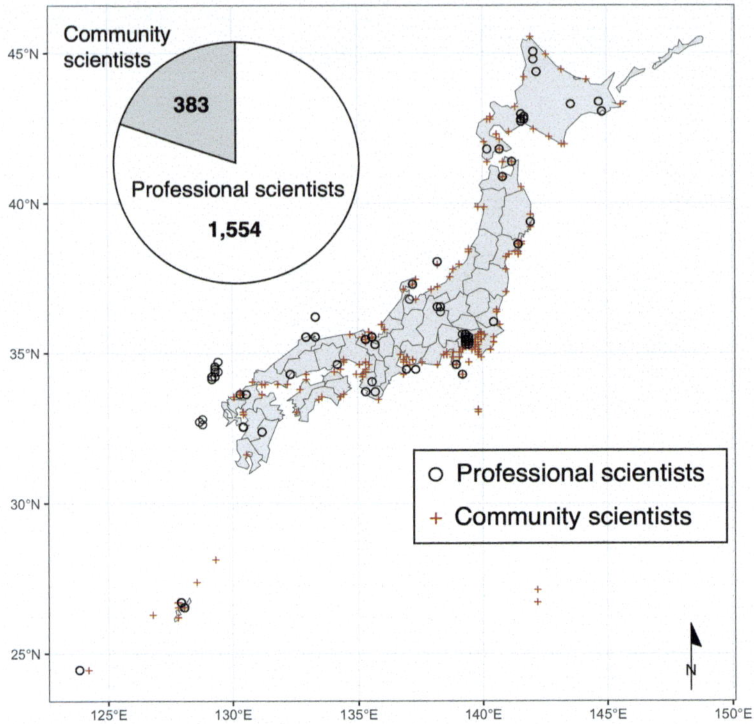


**Fig. 6.3** The sampling locations of ANEMONE by community scientists (crosses) and professional scientists (circles)

ANEMONE DB. However, for conservation purposes, the spatial resolution of the data is reduced to a 10-km granularity.

### *6.4.2 Collaboration with Local Governments and Residents*

ANEMONE has been actively engaged in eDNA monitoring, collaborating with local governments across various regions. One notable case is Minamisanriku Town in Miyagi Prefecture, where ANEMONE conducted eDNA monitoring in partnership with a local government for the first time. Minamisanriku Town, located in the southern part of the Sanriku Coast in northeastern Miyagi Prefecture, boasts a population of approximately 12,000 residents as of 2023. The town is situated along Shizugawa Bay, renowned for its distinctive ria coast scenery, and is deeply involved in marine aquaculture, cultivating species such as coho salmon (*Oncorhynchus kisutch*), pacific oysters, scallops, sea squirt, and seaweed. This coastal area is designated as part of the Sanriku Fukko National Park (Furuta and Seino 2016) and is recognized as a Ramsar wetland. Despite the severe devastation it endured during

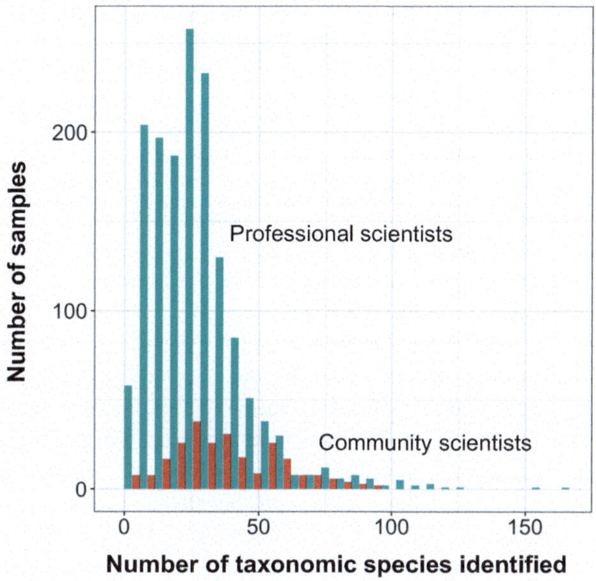


**Fig. 6.4** The numbers of taxonomic species identified from an eDNA survey. Blue and red bars are for professional and community scientists, respectively

the Great East Japan Earthquake in 2011, Minamisanriku Town has been actively pursuing reconstruction efforts, capitalizing on its abundant natural resources. As part of these initiatives, ANEMONE has collaborated with the Minamisanriku Nature Center (Abe and Dazai 2017), a municipal educational and research facility, to conduct ongoing eDNA monitoring in Shizugawa Bay. In addition to biweekly regular observations at designated locations managed by the town, educational activities involving eDNA surveys conducted by local elementary, junior high, and high school students are also being carried out. The survey results are employed to assess the recovery of biodiversity, particularly in relation to the seaweed bed restoration project.

Starting in 2022, ANEMONE initiated a collaboration with Kanagawa Prefecture to conduct eDNA monitoring focused on rivers. Kanagawa Prefecture, a densely populated metropolitan area, is home to approximately nine million residents, making it the second most populous prefecture in Japan after Tokyo. The Sagami River and the Sakawa River, serving as vital water sources for Kanagawa Prefecture, have been the focal points of consistent environmental conservation efforts. As part of these endeavors, ANEMONE established a collaboration, commencing eDNA metabarcoding surveys targeting fish at specified locations since 2020. In 2023, an eDNA survey emphasizing fish in rivers was carried out, involving high school students. The survey results are anticipated to be compiled by the end of the year, and a comprehensive report summarizing the findings will be provided, along with plans for an online presentation meeting to share the results.

### 6.4.3 Collaboration with Shipping Companies for Offshore Observations

Many of the eDNA surveys conducted by ANEMONE have focused on coastal and river environments, and the survey methods used in ANEMONE are designed to be compatible with these ecosystems. While citizen surveys have allowed for an increase in survey points in coastal observations, a similar approach is now expanding to offshore surveys.

ANEMONE's collaboration with the Japanese shipping company Nippon Yusen Kabushiki Kaisha (NYK Line) has paved the way for offshore surveys to be conducted in conjunction with ANEMONE. NYK Line manages a fleet of over 800 vessels, serving more than 350 ports both domestically and internationally, ranking it as Japan's largest shipping company and one of the world's leading shipping companies in terms of fleet size and revenue. Since 2021, ANEMONE has initiated offshore surveys in collaboration with NYK Line and its subsidiary, Kinkai Yusen Kaisha (KYK Line). This collaborative effort involves four key stakeholders: Tohoku University, Hokkaido University, NYK Line, and KYK Line. In 2021, successful DNA detection of 158 fish species was accomplished through surveys conducted along the international route. Starting in 2022, monthly regular observations were launched in the offshore waters of the Sanriku Coast, a critical fishing area.

The realization of offshore surveys along established shipping routes required the development of new technologies. Compared to coastal areas, offshore environments contain lower concentrations of eDNA. While several liters of water sampling can capture numerous species in coastal regions, offshore surveys necessitate the filtration of tens of liters of water. Additionally, commercial ships lack dedicated sampling facilities for scientific surveys. To overcome these challenges, a large amount of seawater sampling was conducted using the water passage of fire hydrants, drawing water from the ship's bottom through pumps. Specialized equipment was devised for onboard filtration of the collected water samples, which were then processed on the ship. These innovations were informed by the expertise gained by Hokkaido University through its prior experience with offshore eDNA surveys conducted on training vessels.

### 6.4.4 Collaborative Mountain Research with Nature Enthusiasts

As previously mentioned, ANEMONE has significantly broadened its scope of target ecosystems through the framework of community science, extending its focus from rivers and coastlines to offshore regions. In 2023, new collaborations have further expanded the purview of ANEMONE's eDNA surveys to encompass mountainous areas that have historically received limited scientific investigation.

Japan, located within the Japanese archipelago, is a notably mountainous country, with mountainous terrain comprising 75% of its land area, featuring over 20 peaks exceeding an elevation of 3000 m. Mountainous regions are particularly susceptible to environmental fluctuations, including rapid warming (Beniston 2003; Kudo 2016). Consequently, conducting ecosystem monitoring in these areas is essential for their conservation. However, mountainous environments pose challenges in terms of accessibility and habitation, necessitating intensified ecological research efforts.

ANEMONE undertakes eDNA surveys in mountainous regions by collaborating with mountain enthusiasts, including hikers and climbers. In 2023, through partnerships with nonprofit organizations, outdoor companies, and nature enthusiasts, ANEMONE conducted eDNA surveys at 90 distinct locations in mountainous areas. This collaborative venture unites mountain enthusiasts with a shared commitment to environmental conservation in the field and scientists keen on conducting biodiversity surveys in remote mountainous terrains. This collaboration has facilitated the execution of challenging biodiversity assessments in these distant mountainous regions. Furthermore, the survey data will be made widely accessible through the ANEMONE DB, thereby contributing to the broader societal understanding. In this respect, this survey stands as a stellar example of the efficacious operation of community science. The scope of targeted organisms encompasses not only fish but also birds and mammals, with plans in place to expand the survey's reach to include amphibians and aquatic insects in the future.

## 6.5 Possible Contribution of eDNA Community Science to the Resolution of Biodiversity Issues

Biodiversity conservation represents one of the most significant societal challenges humanity currently faces. Human activities have expanded to the extent that they can induce changes in the Earth system (Lewis and Maslin 2015). Of the total biomass of mammals on Earth, humans occupy 36% and livestock 60%, while wild mammals account for a mere 4% (Greenspoon et al. 2023). Due to human influence, the rate of species extinction on Earth is occurring at an unprecedented pace, estimated to be 100 times the background extinction rate (Ceballos et al. 2015). The Intergovernmental Science-Policy Platform on Biodiversity and Ecosystem Services (IPBES), in its reports, has estimated that 25% of species are at risk of extinction (IPBES 2019).

The degradation of ecosystems and the loss of biodiversity represent risk factors that threaten the survival of humanity. Ecosystems provide a wide range of ecosystem services. Even when focusing solely on business, many industries, including not only primary industries but also food, construction, and energy, rely on nature. According to the World Economic Forum, it is estimated that every industry has a moderate to high level of dependence on nature, potentially accounting for half of the world's total GDP. To build a sustainable society, there is a need for assessing the state of ecosystems, making future predictions, and establishing sustainable utilization and management practices. In pursuit of these objectives, the species composition of ecosystems stands as the most critical foundational information about ecosystem conditions, and the observation of species composition is indispensable for providing such data.

Biological diversity surveys based on scientific methods, including eDNA analysis, are expected to play an indispensable and significant role in future ecosystem conservation through multiple avenues (Cooper et al. 2007; Dickinson et al. 2010; Bonney et al. 2009). Firstly, community science enables high-frequency surveys across multiple locations. While the development of automated eDNA monitoring devices is underway, conducting frequent eDNA surveys at numerous locations remains a labor-intensive task for professional scientists. On the other hand, local communities are poised to make substantial contributions to biodiversity assessment in their respective regions. In fact, in the case of ANEMONE, surveys conducted by local residents now constitute a significant portion of the overall research effort (Fig. 6.3). Secondly, community science fosters knowledge and civic pride among local residents regarding their regional ecosystems and provides an excellent opportunity for them to engage in ecosystem conservation (Suzuki-Ohno et al. 2023). According to a questionnaire survey of participants, more than 75% of respondents reported an improved understanding of biodiversity, marine environments, and eDNA as a result of their involvement in the project. Approximately 95% of respondents found the eDNA collection work meaningful (Suzuki-Ohno et al. 2023). Thirdly, community-based eDNA monitoring may facilitate consensus-building among local communities regarding sustainable natural resource use. Ecosystems possess diverse functions while also exhibiting regional specificity, which often leads to the concentration of multiple stakeholders within a given region. To ensure the sustainable utilization of ecosystems, it is crucial that stakeholders with the highest interest in sustainability, namely, residents, have access to comprehensive ecosystem information and a significant say in decision-making.

Community science-based eDNA monitoring holds the potential to empower local communities by providing them with valuable ecosystem information. It also serves to enhance the scientific literacy of participants and promotes the sustainable self-governance of regional ecosystems by residents. Within the context of ANEMONE, an initiative initially launched by professional scientists, community science plays an indispensable role in the practical application of eDNA observations for biodiversity conservation. As ANEMONE moves forward, harnessing the power of community science to contribute further to the construction of a sustainable society and expedite its progress, there are challenges that must be addressed. One critical challenge revolves around optimizing the fundraising process to support biodiversity monitoring within ANEMONE. The eDNA observations within ANEMONE are currently sustained through a combination of competitive scientific funding, corporate support, and contributions from local governments, among other sources. To expand biodiversity monitoring efforts, it will be crucial to establish mechanisms that foster collaboration and alignment between the business sector, academia, and local communities. Under the international goal of achieving a Nature-Positive agenda, including the Task Force on Nature-related Financial Disclosures (TNFD), there is a growing trend of active engagement by the business sector, which underscores the significance of fostering connections between local residents conducting biodiversity monitoring and the business sector. The expectation is that this engagement will provide valuable information to the business sector. The establishment of a collaborative framework in which local residents conduct biodiversity monitoring, supported by economic partnerships with the business sector, can simultaneously bolster regional biodiversity monitoring efforts and strengthen corporate commitments to biodiversity conservation. By fostering cooperation among diverse stakeholders, including academia, the business sector, and local communities, we can collectively aim to achieve advanced biodiversity conservation through collaborative efforts, supporting and complementing each other within a framework of synergy.

## References


Abe T, Dazai A (2017) Museum and ecology (28) Nature Center for Regional Society and scientists in Sanriku ria coast: activities before and after Tohoku earthquake and tsunami in Minami-Sanriku. Japan J Ecol 67:67–71

Alerstam T, Hedenström A, Åkesson S (2003) Long-distance migration: evolution and determinants. Oikos 103:247–260

Bauer S, Hoye BJ (2014) Migratory animals couple biodiversity and ecosystem functioning worldwide. Science 344:6179

Beng KC, Corlett RT (2020) Applications of environmental DNA (eDNA) in ecology and conservation: opportunities, challenges and prospects. Biodivers Conserv 29:2089–2121

Beniston M (2003) Climatic change in mountain regions: a review of possible impacts. Clim Chang 59:5–31

Biggs J, Ewald N, Valentini A, Gaboriaud C, Dejean T, Griffiths RA, Foster J, Wilkinson JW, Arnell A, Brotherton P, Williams P, Dunn F (2015) Using eDNA to develop a national citizen science-based monitoring programme for the great crested newt (*Triturus cristatus*). Biol Conserv 183:19–28

Bonney R, Cooper CB, Dickinson J, Kelling S, Phillips T, Rosenberg KV, Shirk J (2009) Citizen science: a developing tool for expanding science knowledge and scientific literacy. Bioscience 59:977–984

Catford JA, Wilson JRU, Pyšek P, Hulme PE, Duncan RP (2022) Addressing context dependence in ecology. Trends Ecol Evol 37:158–170

Ceballos G, Ehrlich PR, Barnosky AD, García A, Pringle RM, Palmer TM (2015) Accelerated modern human-induced species losses: entering the sixth mass extinction. Sci Adv 1:e1400253

Collins RA, Bakker J, Wangensteen OS, Soto AZ, Corrigan L, Sims DW, Genner MJ, Mariani S (2019) Non-specific amplification compromises environmental DNA metabarcoding with COI. Methods Ecol Evol 10:1985–2001

Cooper CB, Dickinson J, Phillips T, Bonney R (2007) Citizen science as a tool for conservation in residential ecosystems. Ecol Soc 12:11. http://www.jstor.org/stable/26267884

Deiner K, Bik HM, Mächler E, Seymour M, Lacoursière-Roussel A, Altermatt F, Creer S, Bista I, Lodge DM, de Vere N, Pfrender ME, Bernatchez L (2017) Environmental DNA metabarcoding: transforming how we survey animal and plant communities. Mol Ecol 26:5872–5895

Dickinson JL, Zuckerberg BJ, Bonter DN (2010) Citizen science as an ecological research tool: challenges and benefits. Annu Rev Ecol Evol Syst 41:149–172

Ficetola GF, Miaud C, Pompanon F, Taberlet P (2008) Species detection using environmental DNA from water samples. Biol Lett 4:423–425

Fish Survey Project Using Environmental DNA (2020) Video manual. https://www.earthwatch.jp/?product=edna

Furuta N, Seino S (2016) Progress and gaps in eco-DRR policy and implementation after the great East Japan earthquake. In: Renaud FG, Sudmeier-Rieux K, Estrella M, Nehren U (eds) Ecosystem-based disaster risk reduction and adaptation in practice. Springer, pp 295–313

Greenspoon L, Krieger E, Sender R, Rosenberg Y, Bar-On YM, Moran U, Antman T, Meiri S, Roll U, Noor E, Milo R (2023) The global biomass of wild mammals. Proc Natl Acad Sci U S A 120:e2204892120

IPBES (2019) Global assessment report on biodiversity and ecosystem services. In: Brondizio ES, Settele J, Díaz S, Ngo HT (eds) IPBES secretariat, Bonn

Komai T, Gotoh RO, Sado T, Miya M (2019) Development of a new set of PCR primers for eDNA metabarcoding decapod crustaceans. Metabarcoding Metagenomics 3:e33835

Kudo G (2016) Landscape structure of flowering phenology in alpine ecosystems: significance of plant–pollinator interactions and evolutionary aspects. In: Kudo G (ed) Structure and function of mountain ecosystems in Japan. Ecological research monographs. Springer, Tokyo

Larson ER, Renshaw MA, Gantz CA, Umek J, Chandra S, Lodge DM, Egan SP (2017) Environmental DNA (eDNA) detects the invasive crayfishes *Orconectes rusticus* and *Pacifactacus leniusculus* in large lakes of North America. Hydrobiologia 800:173–185

Lewis S, Maslin M (2015) Defining the anthropocene. Nature 519:171–180

Minamoto T, Yamanaka H, Takahara T, Honjo MN, Kawabata Z (2012) Surveillance of fish species composition using environmental DNA. Limnology 13:193–197

Minamoto T, Naka T, Moji K, Maruyama A (2016) Techniques for the practical collection of environmental DNA: filter selection, preservation, and extraction. Limnology 17:23–32

Minamoto T, Miya M, Sado T, Seino S, Doi H, Kondoh M, Nakamura K, Takahara T, Yamamoto S, Yamanaka H, Araki H, Iwasaki W, Kasai A, Masuda R, Uchii K (2021) An illustrated manual for environmental DNA research: water sampling guidelines and experimental protocols. Environ DNA 3:8–13

Mittelbach GG, McGill BJ (2019) Community ecology, 2nd edn. Oxford University Press, Oxford

Miya M, Sato Y, Fukunaga T, Sado T, Poulsen JY, Sato K, Minamoto T, Yamamoto S, Yamanaka H, Araki H, Kondoh M, Iwasaki W (2015) MiFish, a set of universal PCR primers for metabarcoding environmental DNA from fishes: detection of more than 230 subtropical marine species. R Soc Open Sci 2:150088

Moisan L, Gravel D, Legagneux P, Gauthier G, Léandri-Breton D, Somveille M, Therrien J, Lamarre J, Bêty J (2023) Scaling migrations to communities: an empirical case of migration network in the Arctic. Front Ecol Evol 10:1077260

Singh NJ, Ecke F, Katzner T, Bagchi S, Sandström P, Hörnfeldt B (2021) Consequences of migratory coupling of predators and prey when mediated by human actions. Divers Distrib 27:1848–1860

Suzuki-Ohno Y, Tanabe AS, Kasai A, Masuda R, Seino S, Dazai A, Suzuki S, Abe T, Kondoh M (2023) Evaluation of community science monitoring with environmental DNA for marine fish species: "fish survey project using environmental DNA". Environ DNA 5:613–623

Taberlet P, Coissac E, Hajibabaei M, Rieseberg LH (2012) Environmental DNA. Mol Ecol 21:1789–1793

Tøttrup AP, Svenningsen L, Rytter M, Lillemark MR, Møller P, Knudsen SW (2021) Citizens in the lab: performance and validation of eDNA results. Citiz Sci 6:1–12

Tsuji S, Takahara T, Doi H, Shibata N, Yamanaka H (2019) The detection of aquatic macroorganisms using environmental DNA analysis—a review of methods for collection, extraction, and detection. Environ DNA 1:99–108

Ushio M, Fukuda H, Inoue T, Makoto K, Kishida O, Sato K, Murata K, Nikaido M, Sado T, Sato Y, Takeshita M, Iwasaki W, Yamanaka H, Kondoh M, Miya M (2017) Environmental DNA enables detection of terrestrial mammals from forest pond water. Mol Ecol Resour 17:e63–e75

Ushio M, Murata K, Sado T, Nishiumi I, Takeshita M, Iwasaki W, Miya M (2018a) Demonstration of the potential of environmental DNA as a tool for the detection of avian species. Sci Rep 8:4493

Ushio M, Murakami H, Masuda R, Sado T, Miya M, Sakurai S, Yamanaka H, Minamoto T, Kondoh M (2018b) Quantitative monitoring of multispecies fish environmental DNA using high-throughput sequencing. Metabarcoding Metagenomics 2:e23297

Ushio M, Hsieh C-H, Masuda R, Deyle ER, Ye H, Chang C-W, Sugihara G, Kondoh M (2018c) Fluctuating interaction network and time-varying stability of a natural fish community. Nature 554:360–363

Valentini A, Taberlet P, Miaud C, Civade R, Herder J, Thomsen PF, Bellemain E, Besnard A, Coissac E, Boyer F, Gaboriaud C, Jean P, Poulet N, Roset N, Copp GH, Geniez P, Pont D, Argillier C, Baudoin JM, Peroux T, Crivelli AJ, Olivier A, Acqueberge M, Le Brun M, Møller PR, Willerslev E, Dejean T (2016) Next-generation monitoring of aquatic biodiversity using environmental DNA metabarcoding. Mol Ecol 25:929–942

Yamamoto S, Masuda R, Sato Y, Sado T, Araki H, Kondoh M, Minamoto T, Miya M (2017) Environmental DNA metabarcoding reveals local fish communities in a species-rich coastal sea. Sci Rep 7:40368

Zhang S, Zhao J, Yao M (2020) A comprehensive and comparative evaluation of primers for metabarcoding eDNA from fish. Methods Ecol Evol 11:1609–1625

Zhu T, Sato Y, Sado T, Miya M, Iwasaki W (2023) MitoFish, MitoAnnotator, and MiFish pipeline: updates in 10 years. Mol Biol Evol 40:msad035

# Chapter 7
# Diverse Motivations in Citizen Science

**Shinichi Takagawa and Yoko Imai**

**Abstract** The "Monitoring Sites 1000 Satoyama project, one of the largest long-term monitoring projects in Japan, has been conducted by more than 2500 citizen surveyors. It is important to understand and manage the motivations of citizen surveyors for continuing the projects. The results of the sociopsychological survey showed a clear difference in motivations for conservation activities between the citizen surveyors and the general public. The sense of obligation (social norms) was the strongest motivation for the general public, and on the other hand, their sense of self-efficacy and attachment to their monitoring site was the strongest motivation for the citizen surveyors. Moreover, the surveyors with at least 5 years of experience had a more diverse types and stronger motivations than the beginner investigators. To sustain the monitoring project over long periods, it is important to devise ways to promote the diverse motivations of citizen surveyors.

**Keywords** Motivation for conservation · Citizen science · Satoyama · Biodiversity monitoring

Citizen science can conduct long-term and wide-area biodiversity monitoring projects that are not possible by individuals. To develop and sustain projects, it is important to understand and manage the motivations of citizen surveyors. The "Monitoring Sites 1000 Satoyama project is one of the largest long-term monitoring projects in Japan, which is managed with the cooperation of the Ministry of the Environment, The Nature Conservation Society of Japan (NACS-J), scientists, and citizen surveyors. This project has been ongoing since 2005, and more than 2500 citizen surveyors are conducting survey of plants, birds, mammals, butterflies, and other indicator species at about 200 survey sites throughout Japan (Fig. 7.1). The survey has

S. Takagawa (✉)
The Nature Conservation Society of Japan, Chuo-Ku, Tokyo, Japan
e-mail: takagawa@nacsj.or.jp

Y. Imai
Department of Urban Engineering, The University of Tokyo, Bunkyo, Tokyo, Japan



Y. Suzuki-Ohno (ed.), *Community Science in Ecology*, Ecological Research Monographs, https://doi.org/10.1007/978-981-97-0304-3_7

revealed national biodiversity trends in agroecosystems and urban ecosystems. The results of this 10-year survey suggested that populations of various common species, such as fireflies, swallows, rabbits, and grassland butterflies, had been declining nationwide (MOE 2019).

In order to investigate what motivates the surveyors to continue the survey, NACS-J and National Institute for Environmental Studies conducted a sociopsychological comparative survey between the 236 surveyors and the 1158 members of the general public citizens without experience of being involved in conservation activities (Imai and Takagawa, unpublished data). The results showed that both groups had different motivations for conservation activities. For the general public, the sense of obligation (social norms) that "everyone should do it" was most strongly related to the willingness to participate in the survey. For citizen surveyors, on the other hand, their sense of self-efficacy and attachment to own monitoring site had the most influence on their motivation to participate in the conservation activities. It was also clear that the surveyors had a greater variety of experiences with nature in their childhood up to 12 years old than the general public. We also found that the years of survey experience affected their motivation to continue the survey. The surveyors with at least 5 years of experience had more diverse and stronger motivations than the beginner surveyors (Fig. 7.2). Sympathy for the activity was the strongest motivation, and all of the other motivations such as self-efficacy, interaction with other generations, and health maintenance were high compared to the beginner surveyors.

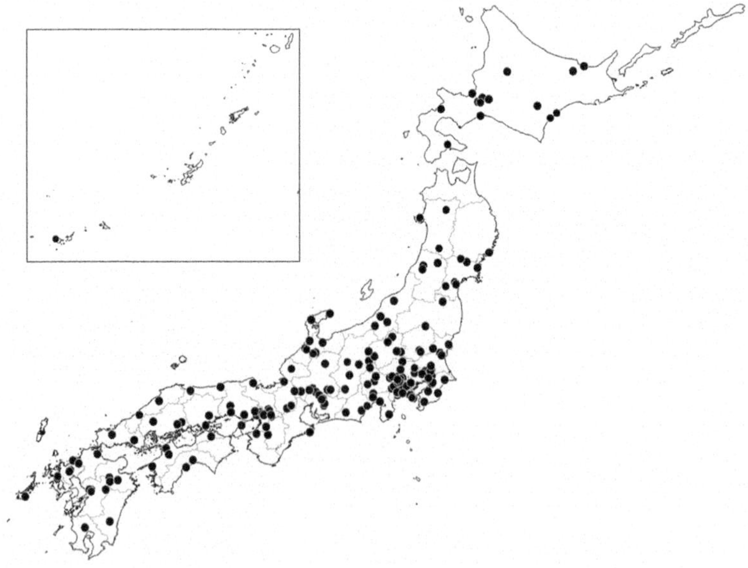

**Fig. 7.1** Location of survey sites of Monitoring Sites 1000 Satoyama project

**Fig. 7.2** Differences in motivation for continuing survey activities by number of years in the survey

These findings suggest that citizen science program for surveying nature fosters a sense of self-efficacy and attachment to the land and promotes motivation to initiate and continue biodiversity conservation activities.

To effectively recruit new surveyors and follow up with existing surveyors for the long-term continuation of the survey, it is important to successfully promote the diverse motivations of citizens.

In the Monitoring Sites 1000 Satoyama project, we have been devising various ways to continue the survey, such as annual exchange meetings across the country, sharing ideas of each survey site for continuing the survey, encouraging new members of the public to participate in the survey, and issuing certificates of appreciation.

## Reference

Minister of Environment (2019) Monitoring sites 1000 Satoyama third phase (FY2005-2017) summary report. Biodiversity Center of Japan, pp 116

# Chapter 8
# Dandelion Survey: Western Japan 2020, Kochi Prefecture. Distribution Survey of Native, Alien and Hybridized Dandelions (*Taraxacum* spp.) through Citizen Participation in Kochi Prefecture

**Kazumi Fujikawa**

**Abstract** A citizen-participatory survey using dandelions, a familiar plant, as an environmental indicator, was initiated in Osaka in the 1970s. Since then, the survey has expanded to the Kinki region and Western Japan, and in Kochi Prefecture, we have been participating in the Dandelion Survey: Western Japan 2010 and have started a distribution survey of Dandelion species in the prefecture in collaboration with citizens. The survey has been continued until the 2020 survey as a two-year survey, with one year of preliminary survey and one year of main survey.

From the results of the survey in Kochi Prefecture in the Dandelion Survey: Western Japan 2020, the native species found in Kochi Prefecture were Kansai Tanpopo (Tanpopo means Japanese name of dandelion), Kushiba Tanpopo, Tsukushi Tanpopo, Yamazato Tanpopo, Kibishiro Tanpopo, Shirobana Tanpopo, and Kibanashiro Tanpopo (yellow flower variety of Shirobana Tanpopo), and the native species presumed to be naturalized in Japan were Tokai Tanpopo, Shinano Tanpopo, and Ezo Tanpopo, and the alien species were Seiyo (Seiyo means Western in Japanese) Tanpopo, Akami Tanpopo and a total of 13 taxonomic groups including native species presumed to be hybrids. The collected samples consisted of 38.2% native species, 60.6% alien species, and 1.1% domestic naturalized species, with alien species accounting for 60% of the samples. The proportion of meshes confirmed to be native involucral bract type alien species in all meshes where alien dandelions were confirmed has doubled in 10 years, and it is thought that the native involucral bract type alien species has doubled in 15 years from the number of collected meshes and samples. In addition, the flowering of Shirobana Tanpopo in the prefecture has been advancing at a rate of 0.9 days over 10 years, and it is presumed that the trend of earlier flowering is due to the long-term rise in temperature during the period when this phenomenon occurs. We have made efforts to promote aware-

---

K. Fujikawa (✉)
The Kochi Prefectural Makino Botanical Garden, Kochi-shi, Kochi, Japan
e-mail: saussure@makino.or.jp



Y. Suzuki-Ohno (ed.), *Community Science in Ecology*, Ecological Research Monographs, https://doi.org/10.1007/978-981-97-0304-3_8

ness by holding exhibitions, training sessions, and issuing newsletters so that citizens can pay attention to and become familiar with the nature around them. After the survey, we published the "Dandelion Survey 2020: Western Japan Kochi Prefecture Report" as a summary of the survey results. This paper is based on this report (Hori Kiyotaka, Tanabe Yuki, Sakamoto Akira, Kokami Yasushi (eds) 2021. Dandelion Survey: Western Japan 2020 Kochi Prefecture Report. pp.63. Public Interest Incorporated Foundation Kochi Prefecture Makino Memorial Foundation. https://www.makino.or.jp/img_data/PAGE_science-new_2_23.pdf?4). We believe that this survey has provided many citizens with an opportunity to think about changes in the natural environment around them. The correspondence table between the Japanese names of dandelion and scientific names is shown in Table 8.1.

**Keywords** Dandelion survey · Kochi · Environmental indicator · Citizen participation

## 8.1 Overview of the Western Japan Dandelion Survey

The "Dandelion Survey" began in Osaka in the 1970s as a method to investigate the state of the environment. Native dandelions, which have been growing naturally in Japan for a long time, thrive in lands that are consistently managed gently, while foreign dandelions, which were introduced to Japan after the Meiji era, have invaded and expanded their distribution in lands significantly altered by development, civil engineering works, and land improvement. By focusing on these differences in habitats, it is thought that by treating dandelions as indicators of the environment, investigating which species of Dandelion is more commonly seen in a certain area can reveal the extent to which the environment of that place has been altered by humans. The reasons why dandelions were chosen as environmental indicators include their ability to accurately reflect the state of the environment, their familiarity to everyone, and the fact that they are rarely mistaken for other plants. In Osaka, many citizens participated in the survey, which started in 1974 as a so-called "citizen participation survey," and since then, the survey has been continued every 5 years. The dandelion survey then spread throughout Japan, including the Kinki region, and surveys were conducted at various scales, such as at the prefectural level, municipal level, or school district level, and data was accumulated. In addition to being viewed as an environmental survey, efforts were also made from the perspective of getting familiar with the nature around us. On a national scale, in 2000, dandelions were treated as indicator species in the Sixth Green National Census "Survey of Familiar Living Things" by the Environmental Agency (now the Ministry of the Environment), and the results were compiled in the 2002 Report on the Survey of Familiar Living Things.

Although surveys were conducted all over the country, the survey years and methods differed by region, and it was not possible to compare the distribution status of each region on the same criteria. In addition, from the late 1990s, a new issue

**Table 8.1** List of scientific names for the Japanese name of dandelion

| Japanese name | Scientific name |
|---|---|
| Kansai Tanpopo | *Taraxacum japonicum* Koidz. |
| Tokai Tanpopo | *Taraxacum platycarpum* Dahlst. var. *longeappendiculatum* (Nakai) Morita |
| Shinano Tanpopo | *Taraxacum platycarpum* Dahlst. subsp. *hondoense* (Nakai ex Koidz.) Morita |
| Kushiba Tanpopo | *Taraxacum pectinatum* Kitam. |
| Ezo Tanpopo | *Taraxacum venustum* H.Koidz. |
| Tsukushi Tanpopo | *Taraxacum kiushianum* H.Koidz. |
| Yamazato Tanpopo | *Taraxacum denudatum* H. Koidz. var. *arakii* (Kitam.) Seriz. |
| Kibishiro Tanpopo | *Taraxacum hideoi* Nakai ex H.Koidz. |
| Shirobana Tanpopo | *Taraxacum albidum* Dahlst. |
| Kibanashiro Tanpopo | *Taraxacum albidum* forma *sulfureum* (H.Koidz.) Kitam. |
| Seiyo Tanpopo | *Taraxacum officinale* Weber ex F.H.Wigg. |
| Akami Tanpopo | *Taraxacum laevigatum* (Willd.) DC. |
| Native involucral bract type alien species | *Taraxacum* sp. (putative hybrid species) |

emerged that was different from when the dandelion survey began in the 1970s, with hybrids between native and foreign species expanding their distribution. Therefore, with the aim of conducting a dandelion survey that includes hybrid dandelions in addition to the traditional native and foreign dandelions, the "Dandelion Survey Kinki 2005 Execution Committee" was organized to conduct the survey throughout the Kinki region using a unified method. The survey was conducted in the seven prefectures of Kinki (Mie, Shiga, Kyoto, Osaka, Hyogo, Nara, and Wakayama) from 2004 to 2005 and achieved significant results (Table 8.1).

## 8.2 Western Japan Dandelion Survey and Makino Botanical Garden

Following the success of the organized dandelion survey in the seven prefectures of Kinki in 2005, the "Dandelion Survey Western Japan 2010" was launched in 2009 as a two-year plan in 19 prefectures in Western Japan. The objectives of conducting the survey in the 19 prefectures of Kinki, Shikoku, Chugoku, and Kyushu (Fukuoka and Saga) are summarized as follows.

1. Record the distribution of different species of dandelions. Keeping a record of what kind of dandelion species are around us can provide various information beyond the environment. For example, some species of dandelions are listed as endangered species in the Red Data Book, but there may be cases where detailed distribution data is not available. We believe that recording the present state of dandelions and passing it on to future generations is of great value.

2. Conduct a dandelion survey in western Japan, which is the distribution area of the Kansai Tanpopo, and understand the current state of the natural environment in these areas from the distribution of native and alien species of dandelions. In areas where surveys have been conducted in the 1970s, it is possible to compare with the past, and even in areas where there is no past data, by comparing with the results of each prefecture, we can obtain materials to think about the characteristics of nature in each region and changes in the natural environment.
3. In the 2005 survey of the Kinki region, it was found that hybrid dandelions have expanded their distribution over a wide range. This is the first time such a dense survey has been conducted throughout western Japan, and it is important to understand the current distribution of hybrid dandelions, including areas where there are few diploid native species.
4. By conducting surveys with more people, each person will pay attention to the nature around them and become interested in changes in the natural environment, and by making it widely known that such large-scale surveys are being conducted, we hope that more people will become interested in the nature around them. We also want to call for "dandelion surveys" as part of environmental education in elementary, junior high, and high schools and foster students who are interested in the natural environment of their local area.
5. By conducting surveys that anyone can easily participate in and thinking about the natural environment in the region, we believe that we can foster exchanges among participants, such as nature conservation groups, museums, nature lovers groups, and plant researchers, and share issues of nature conservation and environmental preservation in each region.

While the native species that is widely distributed in the Kinki region is the Kansai Tanpopo, the species that is widely distributed in the prefecture is the white-flowered dandelion (Shirobana Tanpopo). Also, while the Kansai Tanpopo is a diploid and reproduces sexually, the Shirobana Tanpopo is a polyploid and reproduces asexually. Thus, the native dandelion species and the method of reproduction are different from those in the Kinki region where dandelion surveys have been conducted so far. From this, it was necessary to consider whether the native species of this prefecture, the Shirobana Tanpopo, could become a measure of nature (environmental indicator) instead of the native species of the Kinki region, the Kansai Tanpopo, when conducting a dandelion survey in Kochi Prefecture. Therefore, in Kochi Prefecture, in addition to the objectives of Western Japan 2010, more specific issues were set,

1. To clarify the classification and distribution of dandelion species distributed in Kochi Prefecture.
2. The aim is to clarify the habitat and distribution of the white dandelion and to evaluate its environmental indicator characteristics through comparison with social factors.
3. The purpose and significance of the survey in Kochi Prefecture were clearly stated, such as citizens turning their attention to familiar nature and developing an interest in it.

In the survey of Kochi Prefecture, the Kochi Prefectural Makino Botanical Garden is in charge of the secretariat. They have been calling on the citizens of the prefecture, holding training sessions, identifying the dandelions sent in, and inputting the survey forms. In addition, to smoothly carry out the survey and cover the entire prefecture comprehensively, an executive committee consisting of volunteers from the plant survey has been set up to decide on the survey methods, system, and goals. The participants are mainly those who are active as survey volunteers at the Kochi Prefectural Makino Botanical Garden, citizens interested in the natural environment, and students who are working on it at school. Excluding students, many participants are interested in the natural environment. Also, when it is published in local newspapers, many people send in dandelions growing in their gardens.

## 8.3 Overview of the Survey at Makino Botanical Garden

Kochi Prefecture has been participating since the Western Japan 2010 survey, with the secretariat located at the Kochi Prefectural Makino Botanical Garden, and is working in collaboration with citizens, mainly the survey volunteers of the botanical garden (Fujikawa and Sakamoto 2011). For the 2020 survey, a dandelion survey preparation meeting was held on December 15, 2018, to discuss the survey methods and system and prepare for the establishment of the executive committee. In January 2019, we officially received a request for cooperation from Tomoo Nunotani, representative of the Dandelion Survey Western Japan Executive Committee, and set up a secretariat at the Kochi Prefectural Makino Botanical Garden. Furthermore, we established the Dandelion Survey Kochi Prefecture Executive Committee, mainly consisting of survey volunteers, and confirmed the survey methods, system, and schedule before the start of the survey each year to ensure smooth progress (Fig. 8.1).

The secretariat staff consists of Makino Botanical Garden staff and external experts who have been involved as dandelion survey secretariat, and they have been identifying samples, contacting and coordinating survey participants, grasping the progress of the survey, and planning and holding workshops and exhibitions. The organization of samples and input of survey forms were mainly carried out by short-term employees.

In conducting the survey, we conducted exhibitions and public relations activities to inform people about the dandelion survey and to encourage a wide range of prefectural residents to participate. Also, with the aim of improving the level of survey participants, we held three training sessions in 2019 to learn about the characteristics and habitats of each dandelion growing in Kochi Prefecture on site. We were planning to hold a training session in 2020 as well, but it was cancelled due to the impact of the spread of the new coronavirus infection. Some of the training sessions were held by the Canon Marketing Japan Group "the Furusato (hometown) Project" fund with the aim of passing on a beautiful and verdant hometown for future generations under the Biodiversity Policy.

**Fig. 8.1** The Dandelion Survey Kochi Prefecture Executive Committee (Hori et al. 2021)

As feedback to the survey participants, we sent newsletters each year that included the identification results of the dandelions each person collected and a preliminary report of the survey results. Information about training sessions and newsletters were posted on the Makino Botanical Garden website so that many people could see them. As a summary of the survey results, we published this report, and in place of the survey report meeting that was cancelled to prevent the spread of the new coronavirus infection, we held a mini exhibition of the survey results at the Makino Botanical Garden and set the exhibition period so that many people, including survey participants, could see it over a long period of time.

[Kochi Prefecture Executive Committee]

Kokami Yasushi (Chairman), Enoki Hiroshi, Fukuoka Hiroshi, Hosokawa Kimiko, Ikeda Tomio, Inagaki Noritoshi, Matsumoto Mitsuo, Morimoto Rie, Nakahira Katsuya, Sasaoka Muneo, Takahashi Maki, Tamura Atsuko, Yamanaka Naoaki (titles omitted, in alphabetical order)

[Kochi Prefecture Secretariat]

Fujikawa Kazumi (Representative), Tanabe Yuki (Administration, Sample Identification), Hori Kiyotaka (DNA Analysis, Newsletter & Report Editing), Okabayashi Rika (Exhibition & Public Relations), Kawatani Ritsu (Sample Organization, Data Entry), Sakamoto Akira (External Expert), Kurihara Taeko (External Expert, Data Collection)

[Short-term Employees (Sample Organization, Data Entry)] Shimada Yumi, Hamada Yuka (2019), Mifumi Seto, Kawatani Ritsu (2020)

## 8.4 Survey Method

1. Survey Period Preliminary Survey: February 1, 2019, to May 31, 2019, Main Survey: February 1, 2020, to May 31, 2020
   (a) Note: The survey period for the whole of Western Japan starts from March 1, but in the Kochi Prefecture survey, we started the survey from February 1, one month earlier than other prefectures, to coincide with the flowering period of the Shirobana Tanpopo.
   (b) Note: Due to the spread of the new coronavirus infection, other prefectures plan to extend the survey period by one year to March 1, 2021, to May 31, 2021, but in Kochi Prefecture, we have collected enough samples to compare with the previous survey, so we ended the survey in 2020.
2. Survey System

   The standard regional mesh for Kochi Prefecture is the third mesh (1 km, there are 7018 meshes in a square unit). In the dandelion survey, it is necessary to collect each dandelion growing in each third-order mesh with the accuracy of the third-order mesh in order to analyze the distribution of each dandelion and the proportion of alien dandelions. In past surveys, there has been a tendency for many samples to be collected in Kochi City and its surrounding municipalities where many surveyors reside, and there has been a bias in the collection locations of the samples. Therefore, in order to comprehensively survey the entire prefecture, the executive committee and volunteer surveyors decided on the municipalities and regions to be surveyed and conducted the survey. For areas without a person in charge, the secretariat was responsible for the survey. In the 2020 survey, in order to avoid leaving any areas uninvestigated, a mesh map reflecting the results of the 2019 preliminary survey was created and distributed to the executive committee and volunteer surveyors, and the collection of dandelions in uninvestigated areas and uncollected dandelions was prioritized. At that time, to avoid duplication and waste in the survey areas (meshes), each volunteer surveyor was assigned a specific survey area and asked to conduct the survey.

   Also, in the 2015 survey, although dandelions were collected, a mesh map of dandelions that were not collected in the 2019 survey was created, and efforts were made to collect data that could track the increase and decrease of dandelions compared to the previous survey (Fig. 8.2).

   (a) Note: From the 2015 survey, the location information to be filled in on the survey form was changed from the latitude and longitude/mesh number of the Japanese geodetic system to the latitude and longitude/mesh number of the World Geodetic System.
3. Tools for the survey: The tools used for the survey are as follows. Among them, the survey form and mesh map were distributed to the surveyors. The other tools

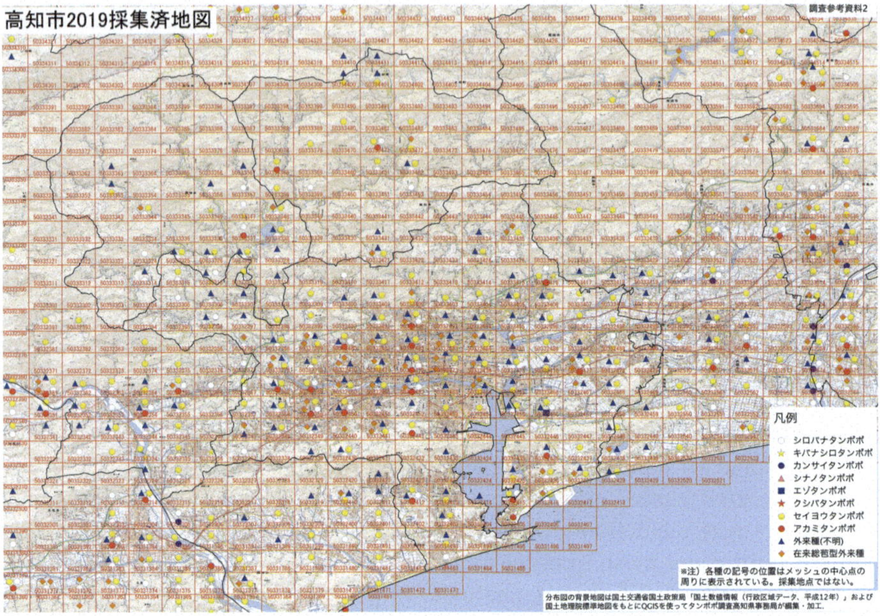


**Fig. 8.2** Mesh map for gap-filling survey (Hori et al. 2021)

were procured by the surveyors themselves. The survey form is a designated form of the Western Japan Executive Committee, and in Kochi Prefecture, to save time when collecting in large quantities and to streamline the organization at the secretariat, a unique envelope-type survey form has been used since the 2010 survey. This time as well, envelope-type survey forms were distributed to executive committee members and volunteer surveyors who collect many samples, and they were used.

- Survey form (designated by the Western Japan Secretariat Fig. 8.3) or envelope-type survey form (Fig. 8.4)
- GPS device, smartphone, tablet, or other tools for knowing location information or mesh map
- Writing utensils
- Clipboard
- Tissue paper
- Cellophane tape

4. Specific Survey Method

   Dandelions do not bloom beautifully on rainy days or early mornings and evenings because they do not get enough sunlight. This is especially true for native dandelions. Therefore, the best time for the survey is from 10 a.m. to 3 p.m. on a sunny day. The specific survey method was as follows.

同じ場所で２種類以上あった場合は、それぞれ別の用紙に記入してください。

## 「タンポポ調査・西日本2020」調査用紙

調査期間：2019年2月1日～5月31日
2020年2月1日～5月31日

| 整理番号 | 記入しない |
|---|---|

**1 調べた日** 20 年 月 日 No. （用紙を複数提出していただく場合）

**2 調べた場所**（できれば番地まで） 住所： 市・町・村

**3 もっと詳しい場所** A・B・Cのどれか一つを記入して

Ⓐ緯度・経度（世界測地系）

| 記入例 | 34 度(°) | 02 分(′) | 24 秒(″) | または | 34.039871 度 |
|---|---|---|---|---|---|
| 北 緯 | 度(°) | 分(′) | 秒(″) | | 度 |
| 東 経 | 度(°) | 分(′) | 秒(″) | | 度 |

緯度・経度を調べた方法 □ホームページ □GPS □携帯電話・スマホ □その他

Ⓑメッシュ番号（世界測地系）
参考：メッシュ地図 http://gonhana.sakura.ne.jp/tanpopo2015/meshmap

| | | | | — | | | — | | |
|---|---|---|---|---|---|---|---|---|---|

Ⓒ目印または地図

（目印の例：○○小学校正門前、△△駅西側100m）

**4 調べた場所の様子** もっとも近いものを１つ選んで○をつけて

A. 林や林のそば B. 池の土手 C. 川の堤防や川原 D. 田畑、果樹園、農道、畦道など E. 神社・寺の境内
F. 公園・校庭・植え込み・グラウンド・団地・家の庭など G. 車道沿い・分離帯 H. 駐車場・造成地
I. その他（ ）

**5 花びらの色は？** もっとも近いものを１つ選んで○をつけて

A. 黄色 B. 白または黄色みがかった白 C. わからない

**6 花（頭花）のかたち** きれいに咲いている花の総苞外片（外側のみどりの部分）はどれに近い？ 番号に○をつけて

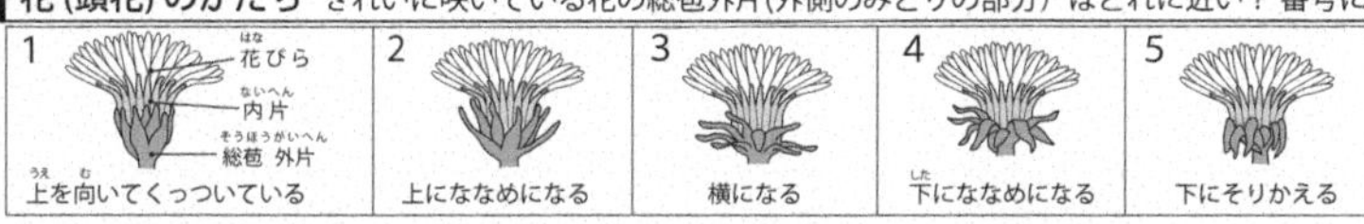


**7 タンポポの種類はどれ？** 番号や記号に○をつけて

1. 花は白色（シロバナタンポポ、キビシロタンポポなど）
2. 花は黄色。総苞外片は上向き（在来種）
3. 花は黄色。総苞外片はそり返る（外来種）

［３に○をした時］タネの色で区別すると、その外来種は次のa～cのどれですか。どれかに○をつけて。
a. セイヨウタンポポ（茶褐色） b. アカミタンポポ（赤褐色） c. タネがないのでわからない

4. わからない

| ここは記入しない | 種名 | | | |
|---|---|---|---|---|
| | 頭花 | 有・無 | 花粉 | 均一・バラバラ |
| | タネ | 有・無 | | ない・判定不能 |

**8 調査をして気がついた点や感想、連絡先を書いてください。** ※この調査に関連した連絡の時にのみ利用します。

感想

調査者の氏名 所属・学校

調査者の住所 〒 電話番号（ ） -

※調査用紙はデータ処理の都合もありますので、各年の６月10日までに各府県の実行委員会宛に送ってください。
※集計したデータは、環境保全や研究目的のため、個人情報を除いて他機関に提供することがあります。

| 整理番号 | 記入しない |
|---|---|

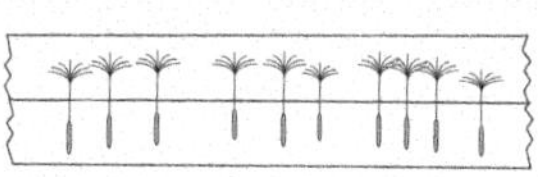

◀タネのはり付け場所
左欄に花と同じ株のタネをセロハンテープではってください。タネがなければかまいません。

必ず、花をティッシュに包んで同封してください。

腐るのでビニールやラップで包まないで!!

**Fig. 8.3** West Japan Office Defined Survey Form (Hori et al. 2021)

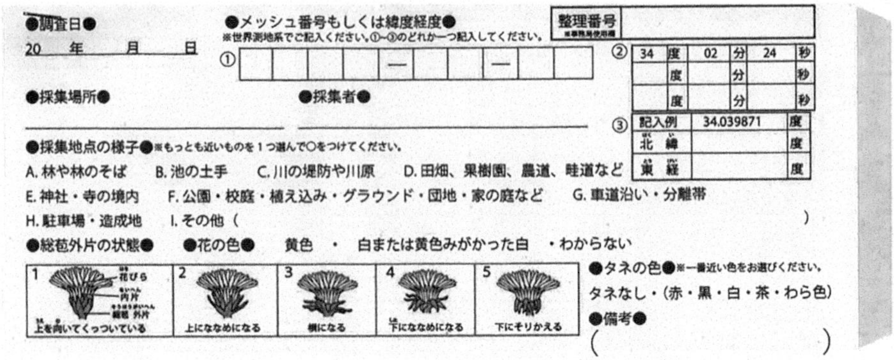
●調査日●
20　年　月　日
●メッシュ番号もしくは緯度経度●
※世界測地系でご記入ください。①~③のどれか一つ記入してください。
整理番号
①
② 34 度 02 分 24 秒
度 分 秒
度 分 秒
●採集場所●
●採集者●
③ 記入例 34.039871 度
北緯 度
東経 度
●採集地点の様子●※もっとも近いものを１つ選んで○をつけてください。
A. 林や林のそば　B. 池の土手　C. 川の堤防や川原　D. 田畑、果樹園、農道、畦道など
E. 神社・寺の境内　F. 公園・校庭・植え込み・グラウンド・団地・家の庭など　G. 車道沿い・分離帯
H. 駐車場・造成地　I. その他（　）
●総苞外片の状態●　●花の色●　黄色　・　白または黄色みがかった白　・わからない
1 花びら 内片 総苞 外片 上を向いてくっついている
2 上にななめになる
3 横になる
4 下にななめになる
5 下にそりかえる
●タネの色●※一番近い色をお選びください。
タネなし・（赤・黒・白・茶・わら色）
●備考●
（　）

**Fig. 8.4** Envelope-type Survey Form (Hori et al. 2021)

(a) Look for a plant that has both flowers and fruit (hereinafter referred to as achene).

- If you can't find one with both, it's okay if it's just blooming.
- Carefully observed to avoid confusion with plants similar to dandelions, such as *Hypochaeris radicata* (Japanese name: Butana), *Ixeris japonica* (Japanese name: Oo-jishibari), and *Sonchus oleraceus* (Jananese name: Nogeshi). Dandelions have all their leaves coming out from the root, spread their leaves in a rosette shape, and the stem that bears the flower does not branch, and only one flower is attached to the tip of the stem, which distinguishes it from similar species.

(b) Filled in the survey form with the survey date, the name of the place where it was collected, location information (latitude and longitude, mesh number, or a map that can identify the place), and the environment of the collection site. The environment of the collection site was classified as follows. (a) near the forest; (b) pond embankment; (c) river embankment; (d) fields, orchards, farm roads, ridges, etc.; (e) shrine/temple; (f) parks, schoolyards, plantings, grounds, housing complexes, home gardens, etc.; (g) along the roadway, median strip; (h) parking lot, reclaimed land; (i) others.

- If a different species was found at the same location, it was recorded on a separate survey form.

(c) Collected a dandelion flower and confirmed the curvature of the involucral bract and recorded it on the survey form.

- Since it is necessary to observe the pollen to distinguish the species of dandelion, a young flower with a lot of pollen (one where the outer flowers are blooming and the central flowers are not) was selected for collection.

(d) Wrapped the flower in tissue paper and put it in a paper envelope.

- Did not use a plastic bag because the flower would rot and the pollen could not be confirmed.

(e) If there were achenes, they were collected and either taped to the survey form with cellophane tape or wrapped in tissue paper and put in a paper envelope.
(f) Any observations, impressions of the survey, and information about the surveyor were recorded on the survey form.
(g) Finally, checked to make sure nothing was omitted.
(h) To prevent the collected flowers and achenes from getting mixed up with the survey sheets, they were stapled together and sent to the prefectural office of the collection site.

    - If not immediately sent to the office, the flowers were stored in a well-ventilated place in a paper bag or similar to prevent them from rotting.

## 8.5 Identification Method

1. Identification method of dandelions indicated on the survey sheet and envelope-type survey form

   In this 2020 survey, as in the previous 2015 survey, we provided check items on the survey sheet about the color of the flower and the shape of the outer bracts of the capitulum and also about the species of dandelions, so that the surveyors could easily identify the samples in the field. The color of the flower is categorized as (a) yellow, (b) white or whitish yellow, (c) unknown, and the shape of the capitulum is categorized as (1) Outer bracts pointing upward and sticking together, (2) slanting upward, (3) horizontal, (4) slanting downward, and (5) curling downward.
2. About the identification of Ezo Tanpopo and Kushiba Tanpopo

   Among the 13 species, the Kochi Prefectural Office, based on its own view, treated those with the curling of the outer bracts being one to three and clearly not corresponding to native species as "alien species with native bract type." Also, among the Kushiba Tanpopo, those that grow in the planting zone and whose leaves are not deeply serrated were treated as Ezo Tanpopo this time.

## 8.6 Data Organization and Analysis

1. Organization of Samples

   The survey forms sent by the surveyors were checked for missing entries, and basically, as the samples arrived, they were assigned an organization number in the order of collector, survey date, city/town/village, and mesh number. The samples were checked, and those judged to be one white-flowered dandelions with prominent angular protrusions, two yellow-flowered dandelions with the curling of the outer bracts being four or five and judged to be Seiyo Tanpopo or

Akami Tanpopo, and three others were separated, and the species of the three others were identified by observing the pollen under a microscope. Missing entries for identification results, flower color, curling, and achene color were filled in with a red pen to distinguish them from the content written by the collector. Note that samples without a capitulum are considered invalid data, but for Akami Tanpopo, even if there is no capitulum, if there are achenes, they are treated as valid data. When checking the samples, if there are achenes in the two Seiyo Tanpopo or Akami Tanpopo, a few achenes were left and separated into another bag and sent to Professor Ito Akira of Osaka City University for hybrid analysis. The analysis results are expected to be compiled in the report for Western Japan.

2. Input of Survey Forms and Data Check

   Once the contents of the survey form were confirmed and the samples were identified, a copy of the survey form was made, and data was entered into the designated Excel sheet set by the West Japan Executive Committee. To eliminate input errors, once a certain amount of data was entered, it was printed out, and proofreading work was carried out. After the survey was completed and all data entry was finished, the location information of all data collection sites, mesh numbers, GPS data, etc., and the species names of the identification results, the curvature of the involucral bracts, and the condition of the pollen were checked and proofread to ensure there were no discrepancies. After the data proofreading was completed, it was sent to the West Japan Executive Committee Headquarters Office by email.

3. Data Aggregation Method

   Among the samples received, those without a single flower and only achenes (excluding only Akami Tanpopo), those that are not dandelions [such similar species as *Hypochaeris radicata* (Japanese name: Butana), *Ixeris japonica* (Japanese name: Oo-jishibari), and *Sonchus oleraceus* (Jananese name: Nogeshi)], and those with unknown collection locations were considered invalid data, and these data were excluded from the aggregation. Also, for the aggregation of collection points (meshes) and the creation of distribution maps, the latitude and longitude location information was converted to standard area mesh codes, and the mesh code location information was converted to decimal latitude and longitude. Data aggregation was done in Excel, and various distribution maps and maps showing the proportion of alien species per mesh (dandelion maps) were created using QGIS. In addition, the number of samples per municipality and the comparison of growth environments were created and included in this report. As of January 20, 2021, the Kochi Prefecture Office has the 2010 survey data, and in the Dandelion Survey/West Japan 2010 Kochi Prefecture Report, those that were considered Oozu Tanpopo (tentative name) were counted as Tokai Tanpopo, and those that were recorded as yellow flowers among the Shirobana Tanpopo were counted as Kibanashiro Tanpopo, and the curvature of the involucral bracts that did not apply to native species among one to three were counted as native involucral bract type alien species. Also, those that were identified but had unknown origins were considered invalid data. In addition, if there were any errors in the collection location data from the sample collectors, the data was corrected.

4. Analysis of the Proportion of Alien Dandelions per Mesh (Dandelion Map)

   The dandelion map is a map that calculates the proportion of alien dandelions in each 1 × 1 km or 2 × 2 km mesh and shows it on a map (Fujikawa and Maeda 2016). The dandelion map has a method of calculating the proportion of alien dandelion samples in the total number of samples collected in each mesh and a method of calculating the proportion of alien dandelion species in the number of native dandelion species and alien dandelion species collected in each mesh. The latter is commonly known as the "Ari-Ari (presence-presence) method" and was proposed by Mr. Ogawa Makoto of the Dandelion Survey/West Japan Office (Curator of the Tokushima Prefectural Museum) (Hotta 2011).

In Kochi Prefecture, a large amount of data has been collected mainly in the central part of Kochi City, where many investigators reside, and there is a bias in the data depending on the region. Also, from the survey method that covers the entire area, where investigators move to the next mesh after collecting dandelions that appeared in each mesh, it is believed that the "Ari-Ari method" will result in a dandelion map closer to the actual situation (Fujikawa and Maeda 2016). Therefore, in this survey as well, we created a dandelion map using the "Ari-Ari method," which calculates the ratio of the number of native dandelions and alien dandelions collected in each third mesh (1 km mesh).

In calculating the ratio of the number of alien dandelion species, the number of native dandelions was counted from the seven species, including the six species of Shirobana Tanpopo, Kibanashiro Tanpopo, Yamazato Tanpopo, Kibishiro Tanpopo, Tsukushi Tanpopo, Kansai Tanpopo, and the Tokai Tanpopo, which are domesticated species but grow in rural environments. For alien dandelions, if neither the Seiyo Tanpopo nor the Akami Tanpopo was collected in the third mesh and if an unidentified alien species was collected, the number of species was counted including the unidentified alien species (those without lean achenes and cannot distinguish between Seiyo Tanpopo and Akami Tanpopo).

### 8.6.1 *Example of Species Count*

**Example 1**

In the case of two Shirobana Tanpopo samples, one Yamazato Tanpopo sample, and three Tokai Tanpopo samples, the native dandelion is counted as "3." Example 2: In the case of three Seiyo Tanpopo samples, one native involucral bract type alien species sample, and two unidentified alien species samples, the alien dandelion is counted as "2." Example 3: In the case of two native involucral bract type alien species samples and two unidentified alien species samples, the alien dandelion is counted as "2." The alien species ratio was calculated using the following formula. Alien species ratio = (number of alien dandelion species/(number of native dandelion species + number of alien dandelion species)) × 100.

## 8.7 Research and Study Results Obtained Through the Dandelion Survey

1. Number of Survey Participants

   In the 2019 preliminary survey, 549 people participated, and the valid samples, excluding samples without flower heads or samples other than dandelions, were collected from 1618 meshes, totaling 4300 samples. In the 2020 main survey, 352 people participated, and valid samples were collected from 2084 meshes, totaling 5791 samples. Combining the preliminary survey and the main survey, there were 762 participants, and a total of 10,091 samples were collected from 2721 meshes. The number of survey participants and the number of samples were the highest ever in the dandelion surveys conducted in Kochi Prefecture (Fig. 8.5).

   Regarding the number of participants, it was significant that there were 294 school participants (5 schools, 301 samples) in the 2019 survey, accounting for 53.6% of all survey participants. The number of general public participants excluding survey volunteers and school participants is accounted for 25.7% of the total survey participants with 141 people. Various factors can be considered for the increase in the number of survey participants this time, one of which could be the placement of a public relations officer in charge of exhibitions and publicity and the ability to do more publicity activities than before. In the 2020 main survey, the number of participants significantly decreased compared to the previous year's survey. This is thought to be due to the spread of the new coronavirus infection, schools being closed during the period of request for self-restraint from going out, and the possibility of participation only became possible just before the end of the survey, limiting it to three schools and 34 people (35 samples).

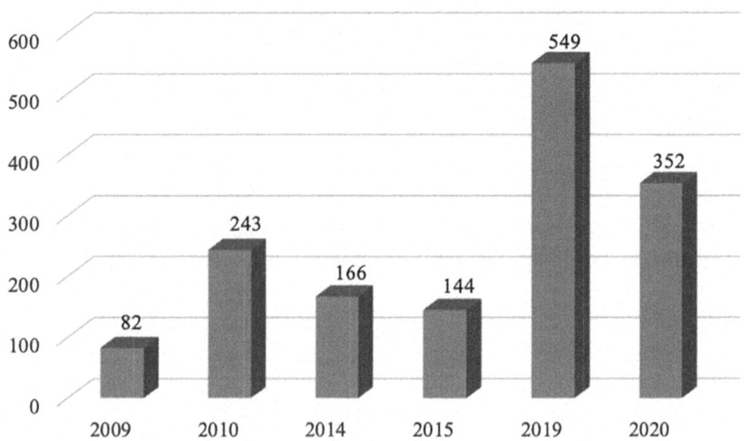


**Fig. 8.5** Number of survey participants (after Hori et al. 2021)

2. Dandelions Growing in Kochi Prefecture

   In the 2020 Dandelion Survey in Kochi Prefecture, it was found that the species as shown in Table 8.2 are distributed in each city and town. The native species that grow naturally are Kansai Tanpopo, Kushiba Tanpopo, Tsukushi Tanpopo, Yamazato Tanpopo, Kibishiro Tanpopo, Shirobana Tanpopo, and Kibanashiro Tanpopo (yellow flower variety of Shirobana Tanpopo); the native species presumed to be naturalized in Japan are Tokai Tanpopo, Shinano Tanpopo, and Ezo Tanpopo; and the alien species are Seiyo Tanpopo, Akami Tanpopo, and native involucral bract type alien species.

3. The Ratio of Species Growing in Kochi Prefecture

   The number of samples of the 13 species of dandelions confirmed in this survey, the number of collection meshes, and the ratio of species sample numbers to all samples are shown in the Table 8.3, showing the transition from the 2010 survey, the 2015 survey, and the 2020 survey. For Tokai Tanpopo, Kibanashiro Tanpopo, and native involucral bract type alien species, the criteria for species determination were adjusted to this survey, and the numbers from the 2010 survey were corrected and used.

   The samples obtained in the 2020 survey were divided into native species (Shirobana Tanpopo, Kibanashiro Tanpopo, Kibishiro Tanpopo, Kushiba Tanpopo, Tsukushi Tanpopo, Yamazato Tanpopo), alien species (Seiyo Tanpopo, Akami Tanpopo, native involucral bract type alien species), and domestic naturalized species (Kansai Tanpopo, Tokai Tanpopo, Shinano Tanpopo, Ezo Tanpopo) and shown in Fig. 8.6. For Kansai Tanpopo, there are naturally occurring (native) populations and domestic naturalized populations, and since most of them are considered to be domestic naturalized, they were treated as domestic naturalized species. The respective ratios were 38.2% for native species, 60.6% for alien species, and 1.1% for domestic naturalized species, and the collected samples resulted in alien species accounting for 60%.

4. Dandelion's Growing Environment

   The growth environments of each dandelion recorded in the survey form are shown in Fig. 8.7. Overall, a high proportion were collected along roadways and medians. This is thought to be due to the fact that the survey was mainly conducted along roads, resulting in a higher proportion of roadways and medians than the actual growth environment

   The Tsukushi Tanpopo, Yamazato Tanpopo that grows at altitudes of 800 m or more in Kochi Prefecture, grows in grasslands (hayfields) near the prefectural border, in shrub forests that have transitioned from grasslands, and in deciduous forests. In this survey, the proportion of forests and near forests accounted for over 80%, which is thought to be due to the fact that we focused on surveying forests that had not been sufficiently surveyed before and that our sampling was biased. The Tokai Tanpopo has a large group along the farm roads in the Fujinokawa area of Shimanto Town (formerly Kubokawa Town), and the proportion of rural environments such as paddy fields and farm roads accounted for 60%. Also, the Kibishiro Tanpopo and Kushiba Tanpopo grow mainly in the hilly areas of the eastern part of the prefecture, and the propor-

**Table 8.2** Dandelions growing in each village, town, and city in Kochi Prefecture (after Hori et al. 2021)

| | Muroto, Chyugei and Aki area | | | | | | | | | Reihoku area | | | | Central area | | | | | | |
|---|---|---|---|---|---|---|---|---|---|---|---|---|---|---|---|---|---|---|---|---|
| | Toyo Town | Muroto City | Yasuda Town | Tano Town | Nahari Town | Kitagawa Village | Umaji Village | Aki City | Geisei Village | Otoyo Town | Motoyama Town | Okawa Village | Tosa Town | Konan City | Kami City | Nankoku City | Kochi City | Ino Town | Hitaka Village | Tosa City |
| Kansai Tanpopo | 1 | | | | | | | | | 3 | | | 5 | 8 | 6 | 34 | 6 | 1 | | 4 |
| Tokai Tanpopo | | | | | | | | | | | | | | | | | | | | |
| Shinano Tanpopo | | | | | | | | | | | | | | | | 3 | 1 | | | |
| Kushiba Tanpopo | | | | | | | | 1 | | 81 | 2 | | 12 | 2 | 60 | | | | | |
| Ezo Tanpopo | | | | | | | | | | 1 | | | | 2 | 3 | | 2 | | | |
| Tsukushi Tanpopo | | | | | | | | | | | | | | | | | | | | |
| Yamazato Tanpopo | | | | | | | | | | | | | | | | | | 2 | | |
| Kibishiro Tanpopo | | | | | | | | | | 26 | | | | | | | | | | |
| Shirobana Tanpopo | 6 | 76 | 15 | 2 | 7 | 17 | 5 | 68 | 27 | 101 | 27 | 12 | 40 | 206 | 318 | 181 | 618 | 170 | 75 | 139 |
| Kibanashiro Tanpopo | | 2 | | | | | | 1 | | 6 | | | 4 | 9 | 21 | 2 | 21 | 8 | 5 | 11 |
| Seiyo Tanpopo | 9 | 51 | 16 | 6 | 8 | 23 | 17 | 56 | 22 | 101 | 61 | 53 | 167 | 143 | 220 | 172 | 544 | 192 | 43 | 84 |
| Akami Tanpopo | 2 | 17 | 4 | 2 | 4 | 2 | | 12 | 6 | 15 | | | | 33 | 54 | 13 | 112 | 22 | 1 | 11 |
| Alien species | 1 | 10 | 1 | | 3 | 5 | 8 | 10 | 1 | 83 | 9 | 18 | 118 | 12 | 118 | 80 | 422 | 108 | 16 | 23 |

| | | | | | | | | | | | | | | | | | | | | |
|---|---|---|---|---|---|---|---|---|---|---|---|---|---|---|---|---|---|---|---|---|
| Native involucral bract type alien species | 4 | 20 | 6 | 3 | 3 | 10 | 1 | 3 | 2 | 33 | 25 | 11 | 148 | 84 | 102 | 34 | 157 | 37 | 4 | 26 |
| Unidentified | | | | | | | | | | | | | | | | | | | | |
| Total number | 23 | 176 | 42 | 13 | 25 | 57 | 32 | 152 | 58 | 450 | 124 | 94 | 494 | 499 | 902 | 519 | 1883 | 540 | 144 | 298 |
| Species number of each area | 5 | 5 | 4 | 4 | 4 | 4 | 3 | 6 | 4 | 9 | 4 | 3 | 6 | 8 | 8 | 7 | 8 | 7 | 5 | 6 |

| | Gohoku area | | | Kouban area | | | | | Hata area | | | | | | | Number of Samples |
|---|---|---|---|---|---|---|---|---|---|---|---|---|---|---|---|---|
| | Sakawa Town | Ochi Town | Niyodogawa Town | Susaki City | Nakatosa Town | Tsuno Town | Yusuhara Town | Shimanto Town | Kuroshio Town | Shimanto City | Mihara Village | Tosashimizu City | Otsuki Town | Sukumo City | | |
| Kansai Tanpopo | | | | | | | 2 | 3 | 6 | 6 | | | | | | 85 |
| Tokai Tanpopo | | | | | | 1 | | 16 | | | | | | | | 17 |
| Shinano Tanpopo | | | | | | 4 | | | | | | | | | | 8 |
| Kushiba Tanpopo | | | | | | | | | | | | | | | | 158 |
| Ezo Tanpopo | | | | | | | | | | | | | | | | 8 |
| Tsukushi Tanpopo | | | | | | 6 | 15 | | | | | | | | | 21 |
| Yamazato Tanpopo | | | | | | 1 | 7 | 5 | | | | | | | | 15 |

(continued)

**Table 8.2** (continued)

| | Gohoku area | | | Kouban area | | | | | Hata area | | | | | | Number of Samples |
|---|---|---|---|---|---|---|---|---|---|---|---|---|---|---|---|
| | Sakawa Town | Ochi Town | Niyodogawa Town | Susaki City | Nakatosa Town | Tsuno Town | Yusuhara Town | Shimanto Town | Kuroshio Town | Shimanto City | Mihara Village | Tosashimizu City | Otsuki Town | Sukumo City | |
| Kibishiro Tanpopo | | | | | | | 2 | | | | | | | | 28 |
| Shirobana Tanpopo | 106 | 82 | 85 | 114 | 83 | 65 | 45 | 309 | 104 | 201 | 14 | 84 | 39 | 32 | 3473 |
| Kibanashiro Tanpopo | 5 | 9 | 2 | 4 | 11 | 2 | 7 | 14 | 4 | 5 | | 1 | 4 | | 158 |
| Seiyo Tanpopo | 85 | 45 | 115 | 63 | 68 | 100 | 106 | 172 | 49 | 98 | 13 | 82 | 43 | 27 | 3054 |
| Akami Tanpopo | 5 | 6 | 13 | 24 | 10 | 13 | 2 | 31 | 9 | 28 | | 10 | 3 | 3 | 467 |
| Alien species | 16 | 14 | 51 | 7 | 17 | 37 | 44 | 105 | 16 | 94 | 13 | 24 | 15 | 20 | 1519 |
| Native involucral bract type alien species | 33 | 18 | 34 | 7 | 32 | 36 | 22 | 57 | 11 | 53 | 6 | 16 | 14 | 26 | 1078 |
| Unidentified | | | | | 2 | | | | | | | | | | 2 |
| Total number | 250 | 174 | 300 | 219 | 223 | 265 | 252 | 712 | 199 | 485 | 46 | 217 | 118 | 108 | 10,091 |
| Species number of each area | 5 | 5 | 5 | 5 | 5 | 9 | 9 | 8 | 6 | 6 | 3 | 5 | 5 | 4 | |

**Table 8.3** The ratio of species sample numbers to all samples (after Hori et al. 2021)

| | Number of samples | | | Mesh number of collected samples | | | Ratio of total sample number (%) | | |
|---|---|---|---|---|---|---|---|---|---|
| Tanpopo species | 2010 survey | 2015 survey | 2020 survey | 2010 survey | 2015 survey | 2020 survey | 2010 survey | 2015 survey | 2020 survey |
| Kansai Tanpopo | 48 | 51 | 85 | 29 | 29 | 37 | 0.8 | 0.7 | 0.8 |
| Tokai Tanpopo | 12 | 14 | 17 | 8 | 8 | 9 | 0.2 | 0.2 | 0.2 |
| Shinano Tanpopo | 9 | 9 | 8 | 3 | 4 | 4 | 0.1 | 0.1 | 0.1 |
| Kushiba Tanpopo | 103 | 163 | 158 | 62 | 82 | 83 | 1.7 | 2.2 | 1.6 |
| Ezo Tanpopo | 0 | 2 | 8 | 1 | 1 | 5 | 0 | 0 | 0.1 |
| Tsukushi Tanpopo | 3 | 11 | 21 | 3 | 7 | 8 | 0 | 0.1 | 0.2 |
| Yamazato Tanpopo | 7 | 7 | 15 | 6 | 6 | 11 | 0.1 | 0.1 | 0.1 |
| Kibishiro Tanpopo | 16 | 19 | 28 | 12 | 12 | 15 | 0.3 | 0.3 | 0.3 |
| Shirobana Tanpopo | 2255 | 2516 | 3473 | 1415 | 1393 | 1767 | 37.4 | 33.3 | 34.4 |
| Kibanashiro Tanpopo | 101 | 120 | 158 | 88 | 84 | 130 | 1.7 | 1.6 | 1.6 |
| Seiyo Tanpopo | 1868 | 2655 | 3054 | 1289 | 1437 | 1871 | 31 | 35.2 | 30.3 |
| Akami Tanpopo | 386 | 639 | 467 | 296 | 325 | 322 | 6.4 | 8.5 | 4.6 |
| Alien species | 861 | 877 | 1519 | 675 | 678 | 906 | 14.3 | 11.6 | 15.1 |
| Native involucral bract type Alien species | 350 | 464 | 1078 | 255 | 376 | 667 | 5.8 | 6.1 | 10.7 |
| Unidentified | 10 | 4 | 2 | 8 | 4 | 1 | 0.2 | 0.1 | 0 |
| Total number | 6029 | 7551 | 10,091 | 2207 | 2304 | 2721 | | | |

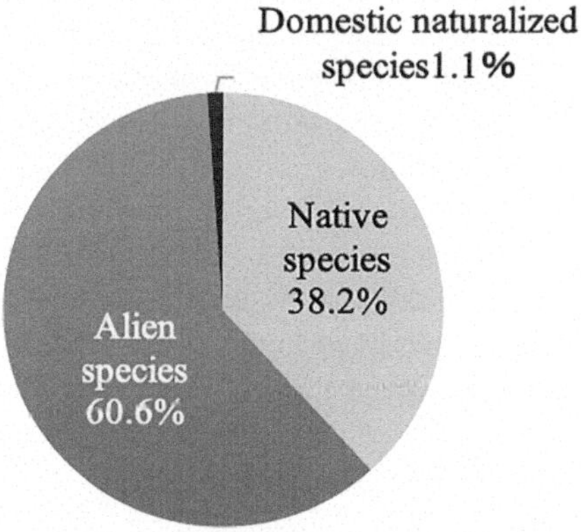


**Fig. 8.6** The ratio of native, alien, and domestic naturalized dandelions (after Hori et al. 2021)

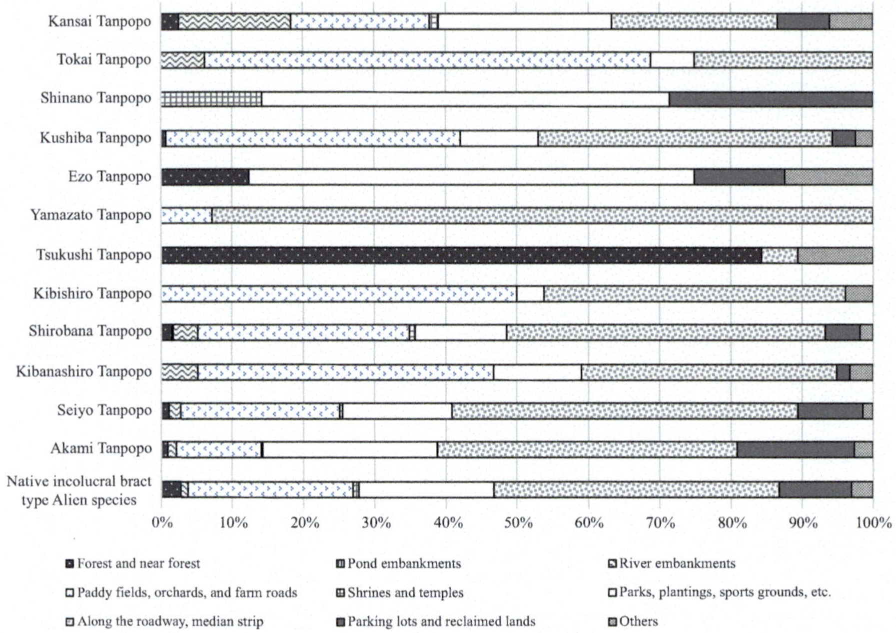


**Fig. 8.7** Proportions of each dandelion's growth environment (after Hori et al. 2021)

tion of rural environments was higher compared to other species. The Kansai Tanpopo had a higher proportion of river embankments compared to other species, which is thought to be due to its growth along the embankments of the Niyodo River, Shimanto River, and Monobe River. These habitats are thought to have a high probability of being brought in from other prefectures when river works and plantings, such as street trees and lawns, are brought in (Fujikawa 2016). The Shinano Tanpopo is thought to have a high proportion because it grows in the plantings at roadside stations, around parking lots, and at shrines. The Shirobana Tanpopo, the most common native dandelion in Kochi Prefecture, has a higher proportion of growth in rural environments such as paddy fields and farm roads compared to alien dandelions, but it also adapts to urban environments, and the most samples were collected along roadways and medians.

There was no significant difference between the Seiyo Tanpopo and the native involucral bract type alien species, but the Akami Tanpopo had a higher proportion of collections in artificially modified and developed lands such as parking lots, reclaimed lands, parks, and plantings compared to the Seiyo Tanpopo and the native involucral bract type alien species. Other surveys in western Japan have also shown that the Akami Tanpopo tends to concentrate in urban areas more than the Seiyo Tanpopo (Dandelion Survey—Western Japan Executive Committee 2016).

5. Transition of Distribution of Foreign Species

For the Seiyo Tanpopo, Akami Tanpopo, and native involucral bract type alien species, we examined the differences in the transition of distribution of the three species based on the data obtained from three surveys over 10 years, from the 2010 survey to the 2020 survey (Tables 8.4 and 8.5).

Looking at the confirmed range of invasive species by municipality, the Seiyo Tanpopo was confirmed in all 34 municipalities in the prefecture in the survey that began in 2010 and continues to the present. The Akami Tanpopo was confirmed in 29 municipalities in the 2010 survey, but was not confirmed in the four municipalities of Umaji Village, Motoyama Town, Okawa Village, and Tosa Town. These four municipalities have not been confirmed in the 2015 survey or in the current survey. For the native involucral bract type alien species, the 2010 survey confirmed them in about half of the 18 municipalities, but the 2015 survey expanded to 33 municipalities excluding Umaji Village, and the 2020 survey confirmed them in all 34 municipalities.

Looking more closely at the distribution of native involucral bract type alien species, the number of confirmed meshes is 255 in the 2010 survey, 376 in the 2015 survey, and 667 in the 2020 survey, increasing 2.6 times over 10 years. Also, looking at the location of the confirmed meshes, it is clear that they have spread throughout the prefecture over 10 years (Fig. 8.8). Looking at

**Table 8.4** Transition of the number of municipalities, meshes, and samples where invasive species were confirmed (after Hori et al. 2021)

| | Number of meshes | | | Number of municipalities | | | Number of samples | | |
|---|---|---|---|---|---|---|---|---|---|
| Species/ survey year | 2010 survey | 2015 survey | 2020 survey | 2010 survey | 2015 survey | 2020 survey | 2010 survey | 2015 survey | 2020 survey |
| Seiyo Tanpopo | 1289 | 1437 | 1871 | 34 | 34 | 34 | 1868 | 2655 | 3054 |
| Akami Tanpopo | 296 | 325 | 322 | 29 | 29 | 29 | 386 | 639 | 467 |
| Native involucral bract type alien species | 255 | 376 | 667 | 18 | 33 | 34 | 350 | 464 | 1078 |
| Total number | 2207 | 2304 | 2721 | 34 | 34 | 34 | 6029 | 7551 | 10,091 |

**Table 8.5** Transition of ratio confirmed meshes and sample proportions for all samples of three alien species (after Hori et al. 2021)

| | Ratio confirmed meshes (%) | | | Ratio of sample numbers (%) | | |
|---|---|---|---|---|---|---|
| Species/survey year | 2010 survey | 2015 survey | 2020 survey | 2010 survey | 2015 survey | 2020 survey |
| Seiyo Tanpopo | 58.4 | 62.4 | 68.8 | 31.0 | 35.2 | 30.3 |
| Akami Tanpopo | 13.4 | 14.1 | 11.8 | 6.4 | 8.5 | 4.6 |
| Native involucral bract type alien species | 11.6 | 16.3 | 24.5 | 5.8 | 6.1 | 10.7 |

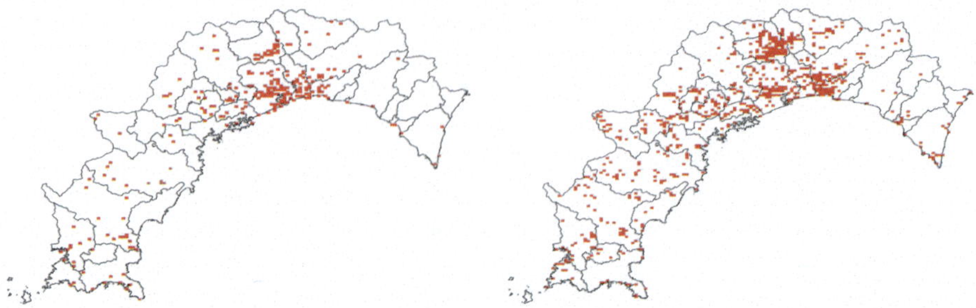

**Fig. 8.8** Distribution change of native involucral bract type alien species. Left: distribution map of the survey in 2010, right: distribution map of survey in 2020 (Hori et al. 2021)

the number of confirmed samples, there were 350 samples in the 2010 survey, 464 samples in the 2015 survey, and 1078 samples in the 2020 survey, increasing 3.1 times over 10 years. As the number of meshes and samples increases with each dandelion survey, looking at the proportion of confirmed meshes of native involucral bract type alien species in all confirmed dandelion meshes, it has increased 2.1 times over 10 years. The proportion of collected samples also increased 1.8 times from 5.8% in the 2010 survey, 6.1% in the 2015 survey, and 10.7% in the 2020 survey. Thus, the native involucral bract type alien species can be seen to have doubled in 15 years from the number of collected meshes and samples. This contrasts with the Akami Tanpopo, which has been almost flat in terms of the number of collected meshes and samples since the 2010 survey.

The native involucral bract type alien species were confirmed in the prefecture during the Kochi Prefecture Plant Survey conducted from 2001 to 2007, and the oldest specimen was collected in Tosa Town on March 28, 2004. Only two specimens, including this one, are cited in the Kochi Prefecture Plant Survey as *Taraxacum* sp. So far, the native involucral bract type alien species have rapidly expanded their distribution range, increased their population, and the number of samples and meshes collected in this survey exceeded the Akami Tanpopo. There is a need to continue to have a strong interest in how it will transition in the future, whether it will exceed the distribution range and population of the Seiyo Tanpopo, and so on.

6. The impact of global warming on the distribution and trends of the Shirobana Tanpopo—Fluctuation of the blooming day of the Shirobana Tanpopo

   (a) Relationship between blooming day changes and temperature

   The blooming day, difference from the average year, comparison with the average year of the winter in western Japan, and climate characteristics (presence or absence of warm/cold winter) from the cumulative years from 1979 to 30 years are shown in Table 8.6. The earliest blooming was on January 7, 2005, and the latest was on April 5, 1984. The average blooming day over 30 years was February 14. The average blooming day for the first 5 years from 1979 to 1984 (no observation value in 1983) was March 12. In contrast, for the most recent 5-year period from 2015 to 2019, it was January 30. The difference in days dur-

**Table 8.6** The blooming day, difference from the average year, comparison with the average year of the winter in western Japan, and climate characteristics in Kochi (after Hori et al. 2021)

| Year | Blooming day | Difference from average year | Comparison with the average year of the winter in western Japan | Climate characteristics (presence or absence of warm/cold winter) |
|---|---|---|---|---|
| 1979 | March 20 | 35 | 1.0 | Warm winter |
| 1980 | March 11 | 26 | −0.6 | |
| 1981 | March 20 | 35 | −2.0 | Cold winter |
| 1982 | February 3 | −11 | −0.8 | |
| 1983 | – | – | −0.4 | |
| 1984 | April 5 | 51 | −2.4 | Cold winter |
| 1985 | March 4 | 19 | −0.7 | |
| 1986 | March 5 | 20 | −2.1 | Cold winter |
| 1987 | February 23 | 9 | 0.5 | Warm winter |
| 1988 | February 25 | 11 | 0.1 | |
| 1989 | January 12 | −33 | 0.9 | Warm winter |
| 1990 | February 21 | 7 | 0.9 | Warm winter |
| 1991 | March 6 | 21 | −0.1 | |
| 1992 | February 26 | 12 | 0.9 | Warm winter |
| 1993 | February 10 | −4 | 1.0 | Warm winter |
| 1994 | March 2 | 17 | 0.0 | |
| 1995 | January 11 | −34 | 0.2 | |
| 1996 | March 6 | 21 | −1.0 | Cold winter |
| 1997 | February 26 | 12 | −0.2 | |
| 1998 | March 4 | 19 | 1.1 | Warm winter |
| 1999 | March 3 | 18 | 0.6 | |
| 2000 | January 19 | −26 | −0.2 | |
| 2001 | February 6 | −8 | 0.2 | |
| 2002 | February 19 | 5 | 0.5 | |
| 2003 | February 12 | −2 | 0.1 | |
| 2004 | February 16 | 2 | 0.4 | |
| 2005 | January 7 | −38 | 0.3 | |
| 2006 | February 9 | −5 | −0.8 | Warm winter |
| 2007 | January 10 | −35 | 1.4 | Warm winter |
| 2008 | February 7 | −7 | 0.1 | |
| 2009 | January 23 | −22 | 0.9 | Warm winter |
| 2010 | February 2 | −12 | 0.7 | |
| 2011 | February 16 | 2 | −0.4 | |
| 2012 | February 23 | 9 | −0.8 | Cold winter |
| 2013 | February 6 | −8 | −0.7 | |
| 2014 | February 12 | −2 | −0.1 | |
| 2015 | February 12 | −2 | −0.2 | |
| 2016 | January 20 | −25 | 1.0 | Warm winter |
| 2017 | February 3 | −11 | 0.8 | |

(continued)

**Table 8.6** (continued)

| Year | Blooming day | Difference from average year | Comparison with the average year of the winter in western Japan | Climate characteristics (presence or absence of warm/cold winter) |
|---|---|---|---|---|
| 2018 | February 7 | −7 | −1.2 | Cold winter |
| 2019 | January 15 | −30 | 1.3 | Warm winter |
| 2020 | January 23 | −22 | 2.0 | Warm winter |

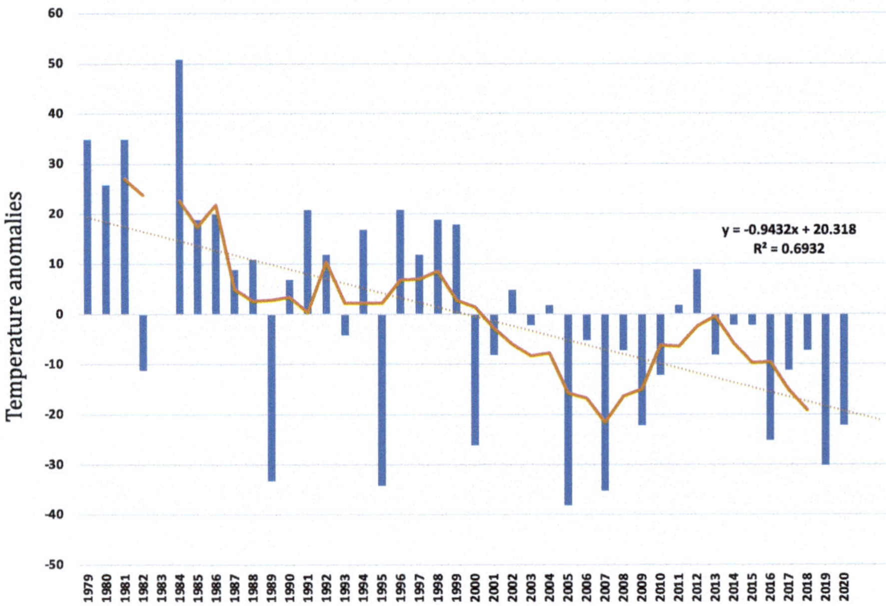


**Fig. 8.9** The average difference and long-term trend of the blooming day of the Shirobana Tanpopo in Kochi Prefecture (after Hori et al. 2021)

ing this period was 42 days. The average difference and long-term trend of the blooming day of the Shirobana Tanpopo in Kochi Prefecture are shown in Fig. 8.9. Because there is variation in the blooming day, the trend was calculated from the regression line of the five-year moving average of the deviation, resulting in a trend = −0.94 (days/10 years) and a determination coefficient of 0.69. The trend of the annual average temperature in Kochi Prefecture (1886–2018) is 1.5 (°C/100 years), and the winter season (December to February) when the Shirobana Tanpopo begins to bloom is 1.2 (°C/100 years) (Kochi Local Meteorological Observatory 2019). From these facts, it was estimated that the blooming of the Shirobana Tanpopo in the prefecture is advancing at a rate of 0.9 days per 10 years, and the trend of earlier blooming is influenced by the long-term temperature rise during the period when this phenomenon occurs.

(b) Relationship between blooming day and temperature

The blooming of dandelions is particularly related to temperature (Masuda et al. 1999; Morita and Kato 2013). Although the blooming day of the Shirobana

Tanpopo did not necessarily become earlier in warm winters or later in cold winters, such a trend was observed. The blooming day of each year showed a negative correlation (the higher the temperature, the earlier the blooming day) with the average temperature of December of the previous year and January to March of the blooming year. The correlation coefficient was largest in January, at −0.56. From previous studies, "Dandelion" is 2, three many regions have their flowering days in February, and a high correlation with the average temperature of February and March has been reported (Masuda et al. 1999). However, in Kochi Prefecture, there are many years when the flowering day is in February (21 times), and it was found that the correlation with the average temperature of the previous month, January, is the highest.

No clear trend was observed between the temperature index and the flowering day of the Shirobana Tanpopo. The number of days when the accumulated temperature just before flowering exceeded 100 °C was 26–32 days in the 5 years when flowering was early (1989, 1995, 2005, 2007, 2019), and 21–54 days in the 5 years when flowering was late (1979, 1980, 1981, 1984, 1996).

## 8.8 Advantages and Disadvantages of the Western Japan Dandelion Survey

### *8.8.1 Disadvantages*

- In Kochi Prefecture, the Makino Botanical Garden serves as the secretariat, processing survey forms sent by citizens and publishing the results. However, not all prefectures have natural history museums or botanical gardens, and it is not possible to handle this as a task. It is a considerable amount of work, and while some prefectures participate in the survey, they may not be able to compile the results.
- In regions where diploid dandelions, including those in Kochi Prefecture, do not grow much, dandelions are not environmental indicators.

### *8.8.2 Advantages*

- Large-scale surveys are conducted continuously, allowing us to feel changes in the natural environment and understand regional characteristics.
- As a citizen-participation survey, the method is established, making it possible to utilize the "Dandelion Survey" for environmental education (however, this depends on the prefecture).

## 8.9 Lessons Learned from the Western Japan Dandelion Survey

The survey targets familiar plants, and the method is simple, so more citizens are paying attention to the natural environment. There is also participation at the school level, providing opportunities for environmental education.

On the other hand, not all prefectures have facilities with experts in the field of natural science, and there is a regional disparity in conducting citizen science. Observing in the field is fundamental, and to solve the mysteries and puzzles in nature together, guidance and compilation by experts and their publication are essential. It is desirable to have natural science museums or botanical gardens in each prefecture.

## References

Dandelion Survey—Western Japan Executive Committee (ed) (2016) Dandelion Survey—Western Japan 2015. Survey Report. Dandelion Survey—Western Japan Executive Committee, Osaka, 174pp (in Japanese)

Fujikawa K (2016) Species growing in Kochi Prefecture and their distribution. Kansai Dandelion. In: Kochi Prefecture Dandelion Survey Executive Committee & Kochi Prefecture Makino Memorial Foundation (eds) Dandelion survey—Western Japan 2015. Kochi Prefecture Report, p 9. (in Japanese)

Fujikawa K, Maeda A (2016) Dandelion map. In: Kochi Prefecture Dandelion Survey Executive Committee & Kochi Prefecture Makino Memorial Foundation (eds) Dandelion survey—Western Japan 2015. Kochi Prefecture Report. Kochi Prefecture Makino Memorial Foundation, pp 19–21. (in Japanese)

Fujikawa K, Sakamoto A (2011) Dandelion survey—Western Japan 2010. Kochi Prefecture Report. Kochi Prefecture Executive Committee, pp 66. (in Japanese)

Hori K, Tanabe Y, Sakamoto A, Kokami Y (eds) (2021) Dandelion Survey—Western Japan 2020. Kochi Prefecture Report. Kochi Prefecture Makino Memorial Foundation, pp 63. https://www.makino.or.jp/img_data/PAGE_science-new_2_23.pdf?4. (in Japanese)

Hotta K (2011) The relationship between dandelion species and population dynamics and density. In: Fujikawa K, Sakamoto A (eds) Dandelion survey—Western Japan 2010. Kochi Prefecture Report. Dandelion survey—Western Japan 2010. Kochi Prefecture Executive Committee, pp 33–45. (in Japanese)

Kochi Local Meteorological Observatory (2019) Long-term changes in temperature in Kochi Prefecture. https://www.jma-net.go.jp/kochi/kouchi_kikouhendou/4-4-2/kouchi_kikou_t.html. Accessed 8 Dec 2020. (in Japanese)

Masuda K, Yoshino M, Park E (1999) The impact and detection of global warming by biological seasons. Global Environ 4(1):91–103. (in Japanese)

Morita S, Kato H (2013) Meteorological factors affecting dandelion flowering. Bull Inst Nat Sci Nihon Univ 48:83–93. (in Japanese)

# Chapter 9
# Citizen Science and Natural History Museums in Japan

**So Ishida**

**Abstract** In Japan, organized citizen science (public participation in scientific research) became widespread during the 1970s. Since then, natural history museums (NHMs) have played a significant role in supporting research activities by citizens and in promoting their own projects. In particular, the Friends of the Museum (FoM), a group of people who use museums as a base for learning and supporting museum activities, has functioned as a hub for amateur naturalist activities and has supported citizen science at both individual and organizational levels. Since the 1980s, the number of prefectural and municipal museums has increased, and citizen science has grown in close collaboration with local communities. The development of digital and internet technologies since the 1990s has revolutionized citizen science, and projects that utilize them are also being carried out in NHMs. The methods used, the results obtained, and the philosophy of citizen science embody the same regional focus and research and educational activities as NHMs, and many participants have been involved in collaborations. Future issues include opening data, securing funding, and fostering the next generation. The last in particular is a mission that NHMs are expected to fulfill.

**Keywords** Naturalist · Local community · Partnership · Ecological monitoring · Museum education · Online database · Social media

## 9.1 The Dawn of Citizen Science in Japan and Museums

Citizen science (defined in this chapter principally as public participation in scientific research) has a long history in Japan, but did not become widespread in organized form until the early 1970s (Kobori et al. 2016). Urban development was advancing along with the country's rapid economic growth, and living things that had been familiar to residents were gradually disappearing. Feeling a sense of crisis,

S. Ishida (✉)
Osaka Museum of Natural History, Higashisumiyoshi-ku, Osaka, Japan
e-mail: iso@omnh.jp



Y. Suzuki-Ohno (ed.), *Community Science in Ecology*, Ecological Research Monographs, https://doi.org/10.1007/978-981-97-0304-3_9

people surveyed the distribution of organisms in specific areas to try and show that the loss of the natural environment was being accompanied by a drop in the number of organisms. For example, in 1971, the Nature Conservation Society of Hyogo Prefecture called on its members, together with school teachers and students, to provide information on snails, paper wasps, and Japanese silver grass (The Nature Conservation Society of Hyogo Prefecture 1971). The Society received 2180 submissions, mainly from southern Hyogo and northern Osaka prefectures, and produced a map showing the degree of destruction of nature in that region as defined by the number of species that occur there (Fig. 9.1; The Nature Conservation Society of Hyogo Prefecture 1972). Although this example was somewhat unsophisticated in that it did not take into account ecological characteristics of each species that are thought to affect their distribution, the number of submissions indicates that many urban residents were aware of the problem. It can thus be seen as a pioneering example of local awareness being mobilized through a survey.

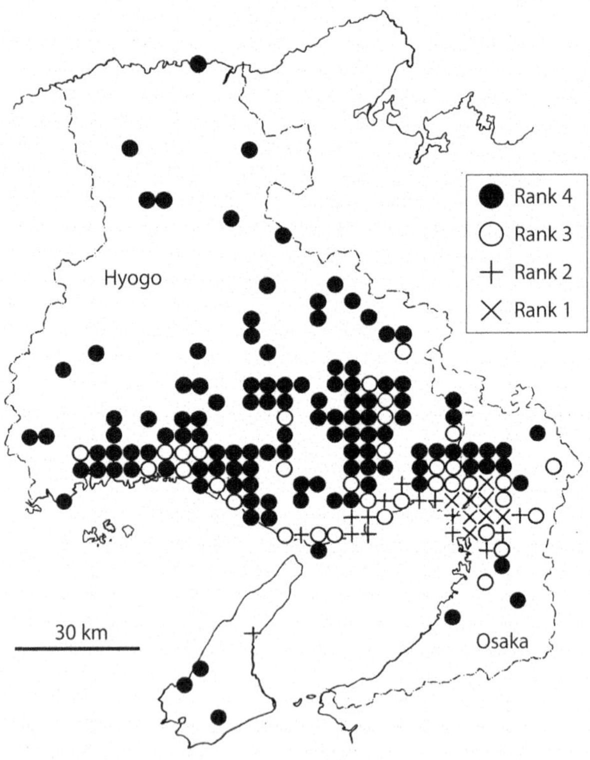


**Fig. 9.1** Result of the land snail occurrence survey conducted by the Nature Conservation Society of Hyogo Prefecture in 1971. Rank denotes the occurrence frequency calculated by the number of presence reports (presence reports per total submission). Rank 4: 60–100%; 3: 30–59%; 2: 0–29% (total submissions <5 and no presence report categorized here); 1: 0% (total submissions 5–10) or not exceeding 10% (total more than 11) (NCSHP 1972: Fig. 3 by courtesy of the Society, partially modified)

Among similar efforts, the Dandelion Survey ('Tampopo Chosa' in Japanese) was an example that succeeded in achieving greater scientific precision. While native dandelions are distributed throughout Japan, the introduced European Dandelion *Taraxacum officinale* was only confirmed in the early twentieth century (Makino 1904). *T. officinale* became common in many areas in the 1920s and 1930s, but did not expand its distribution into urban areas until the 1960s (Hotta 1975a, b). Noting that the European Dandelion was becoming widespread in urban developed areas, a group of citizens in the 1970s began to survey the distributions of both native and non-native species in the Kyoto-Osaka-Kobe region. Mitsuru Hotta of Kyoto University was involved in the leadership and analysis of these surveys, which used bioindicators to depict the degree of urbanization (Fig. 9.2; Hotta 1977). Dandelions are familiar to many Japanese people and easy to find, and it is relatively simple to distinguish the native species from the European one by its involucral scales. These factors suited this study to a citizen-science approach.

The survey subsequently spread to other regions (Ogawa 1979; Ogawa and Mototani 1985; Hamaguchi 1989; The Executive Committee of Dandelion Survey in West Japan 2016). In many areas, natural history museums served as bases for surveys. In the Kanto region, the Hiratsuka City Museum and the Kawasaki Municipal Science Museum were responsible for organizing regional surveys together with high school biology clubs (Hamaguchi 1989). In western Japan, surveys have since continued over a broader area, with regional NHMs hosting nine of the 14 official prefectural participants in the 2020 survey. There are three reasons why NHMs have become the bases for this project. The first is that their existing mission is to collect and accumulate information on local natural history, the second is that they are open on a daily basis for amateur naturalists to meet, and the third is

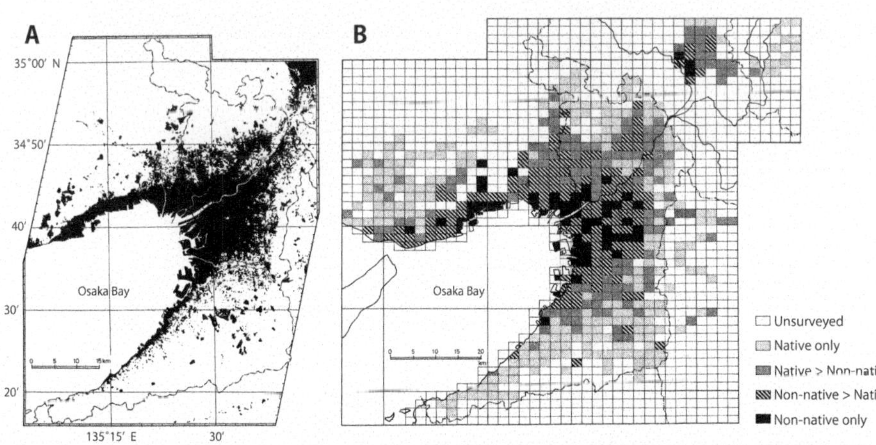


**Fig. 9.2** Results of the dandelion surveys conducted in Kyoto-Osaka-Kobe region in 1973–1976. (**a**) Urbanized area estimated by Landsat 1 remote sensing data shot in 1972; (**b**) Occurrence ratio of native (*Taraxacum japonicum*) and non-native (*T. officinale*) dandelions at each mesh (Hotta 1977: Figs. 12 and 11, partially modified)

that survey activities of this kind are examples of the social education that NHMs aim for. The second point is a great advantage in terms of assembling volunteers who are skilled in searching for and identifying organisms. All three factors were cited by Sforzi et al. (2018).

While it is difficult to say for sure whether the Dandelion survey influenced subsequent NHM activities, it is true that, since the 1980s, various citizen science surveys have been organized primarily by NHMs themselves. One example is a marking survey of the chestnut tiger butterfly *Parantica sita niphonica* ("Asagi-madara" in Japanese, Fig. 9.3), a species in the family Danaidae. Based on its seasonal appearances in Okinawa and Kagoshima, it was hypothesized that the chestnut tiger migrates long distances across the Japanese Archipelago. One way to verify this is to employ the same mark-recapture method used in migratory bird surveys. Although large numbers of markings are needed to achieve even a few recapture cases, citizen science could make this possible. Hiroshi Tanaka of the Kagoshima Entomological Club and Haruo Fukuda of the Kagoshima Prefectural Museum cooperated in calling for participation in the survey, and 4000

**Fig. 9.3** Chestnut tiger *Parantica sita niphonica*. Marking was made by felt-tip pen on the wings (Photo by Itaru Kanazawa)

individuals were marked during the 3 years from 1980 to 1982. In 1981, three marked individuals were recaptured at between 400 and 1200 km from Kagoshima, greatly advancing the long-distance migration hypothesis (Fig. 9.4; Hiura 1983a; Fukuda 1991; Kano 2003). After hearing about these results from Fukuda, Isamu Hiura of the Osaka Museum of Natural History established the "Association for Studying the Chestnut Tiger" in 1983 and invited the members of the Friends of the Museum society to participate in marking surveys (Hiura 1983b). Hiura was extremely skilled at encouraging and organizing amateurs (Takibata 2003), but regrettably, he passed away suddenly 4 months after starting the association. The survey nevertheless expanded into a nationwide network, resulting in the marking of more than 10,000 individuals each year. According to the international marking survey network, the migration route has been clarified as northward from Taiwan, Hong Kong, and southeastern China to Kanto and Hokuriku in Japan during spring and summer and then south again in autumn (Fig. 9.5; Miyatake et al. 2003, Kanazawa et al. 2017).

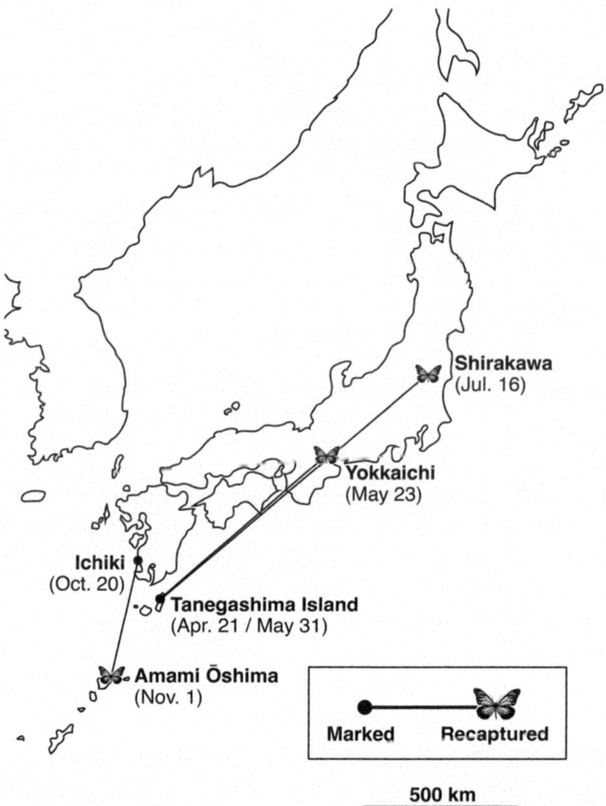


**Fig. 9.4** Three recapture records of chestnut tigers marked at Kagoshima Prefecture in 1981 (drawn from the data in Fukuda 1991)

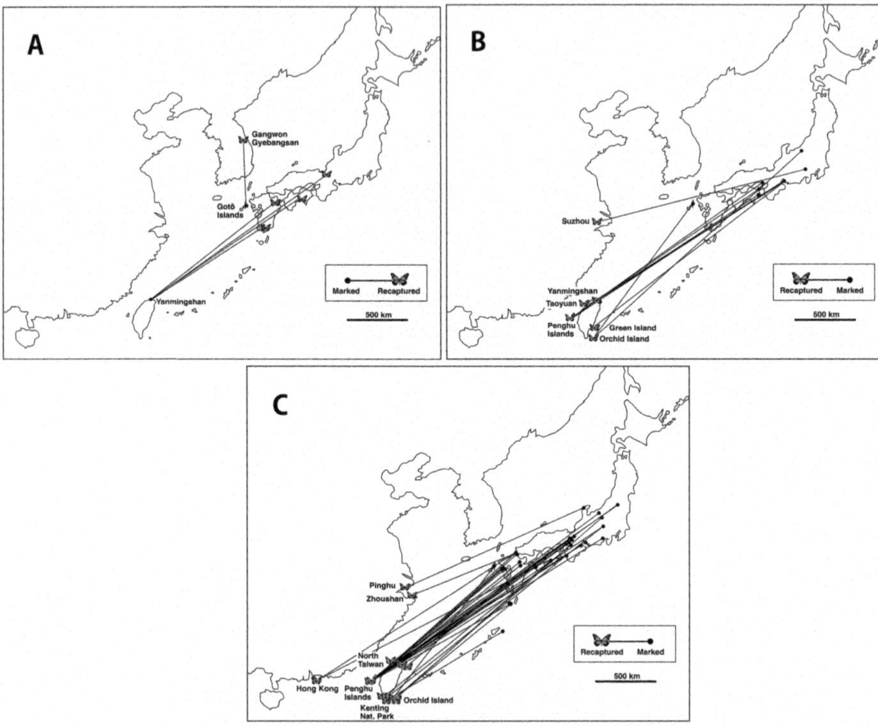


**Fig. 9.5** Mark-recapture records of chestnut tiger migration across countries during 2000–2016. (**a**) Northbound migration by males and females; (**b**) Southbound by females; (**c**) Southbound by males. Note that each line does not necessarily reflect the actual trajectory during their migration (same in Fig. 9.4). (redrawn from Kanazawa et al. 2017: p. 4–5, by courtesy of the authors)

## 9.2 Friends of the Museum and Citizen Science

Friends of the Museum ("Tomo no kai" in Japanese), or groups of people who use museums as a base for learning and for supporting museum activities, are organized throughout Japan. Although FoMs have long been found in other kinds of Japanese museum (Japanese Association of Museums 1963, 1979), it was not until the 1980s that an increasing percentage of natural history museums began to have them (Sakuma 2015). Although the organizational forms and goals of FoMs vary slightly from museum to museum, according to a 2013 survey, about 30% of natural history museums in Japan had one. Of these, half stated that the purpose of their FoM is to promote independent learning activities by having members use the museum or participate in organized educational programs there (Hamada 2015). The friends of natural history museums include not only children who love living creatures and fossils but also amateur naturalists who are experts in a particular taxon or subject. Amateur naturalists receive support from the museum through access to specimens, bibliographic information, research facilities, and a place to present their findings, while in return, they give the museum donations of observational information and specimens. They also voluntarily assist the museum's collection sorting/cataloging and educational activities. It is quite natural that research activities develop through such relationships, and countless research records are created by individual FoM members. Since no demand is made on them for short-term results, their research is often based on long-term observations (Suzuki et al. 2010). In addition, taxonomy everywhere is largely supported by amateurs, especially in insects (e. g., Hopkins and Freckleton 2002), and this is also true in Japan. It is not uncommon for the FoM itself to be the center of their activities.

While many cases of citizen science involve individuals, FoMs also conduct group research, as in the case of the chestnut tiger survey. In the dandelion survey in western Japan, some museums have called on their FoMs to collect data. In another example, the Osaka Museum of Natural History has been conducting projects to survey local biota over a span of several years in cooperation with its FoM (Fig. 9.6). Each project focuses on water systems, urban areas, or introduced species and comprehensively surveys plants, insects, the benthos, fish, birds, amphibians, reptiles, mammals, water quality, and other factors. The data are published (e. g., Shiyake, and Beetle Team of Yamatogawa River Research Group (Project Y, Osaka Museum of Natural History) 2008; Shiga, and Aquatic Plant Research Team of Yamatogawa River Research Group (Project Y, Osaka Museum of Natural History) 2008; Nakaguchi et al. 2013; Shiyake et al. 2020; Ishida 2020a). In citizen science, it is important to feed back the results to participants and society (Silvertown 2009). In these projects, special exhibitions were held at the end of the surveys to present the results (Fig. 9.6). This is a unique kind of feedback that can only be done at a museum.

**Fig. 9.6** "Project A (Alien species survey in Osaka and adjacent areas)" conducted by the Osaka Museum of Natural History and the society of friends of museum during 2015–2019. **Left**: Training the participants to search for and identify introduced fishes. After that they searched for the target species around residential areas; **Right**: Special exhibition held in 2020 at the project completion, describing the distribution of introduced species and the project activities

## 9.3 Museums as Hubs Connecting Local Communities

Changes in the museum community in Japan during the 1980s and 1990s included an increase in the number of local institutions. According to the Japanese Association of Museums (2009), the number of national museums changed little between 1981 and 1999, while that of prefectural museums grew from 79 to 132 and municipal museums from 190 to 417. Regardless of their field, these local museums have made it their mission to collaborate with their local community. In the case of natural history museums, they function as hubs for collecting and disseminating information on nature at the center of the community. In this context, a movement has arisen to build a network among museums. While individual NHMs offer a variety of citizen science activities, by working together they can expand their activities to cover a wider range of areas. The Dandelion survey in western Japan is, of course, one example of this.

Another good example of leveraging a network is the Hokkaido Flowerthon (Hokkaido Flowerthon 2022). This survey has been conducted every 5 years since 1997. For 2 days in mid-June, volunteers report the species and locations of wild flowers blooming in Hokkaido, and a distribution map is then generated (Fig. 9.7). Between 2000 and 3000 people participate each year, and approximately 1100 species have been confirmed. Consecutive surveys have described transitions in floral phenology and the distribution of various species (Fig. 9.8; Watanabe and Niwa 2022). The survey comprises 950 meshes, each of about 100 km$^2$, and citizens' groups and 18 NHMs in Hokkaido serve as advisors and provide support to the volunteers. Several NHMs have organized off-site events as part of the survey. The Flowerthon was originally led by local circle members who were knowledgeable about plants, but their number is decreasing due to aging and other factors. NHM events have attracted individual community residents, and the museum has thus become a place where they can be supported (O. Watanabe, pers. comm.) A strong network among museums in Hokkaido has been established over the years,

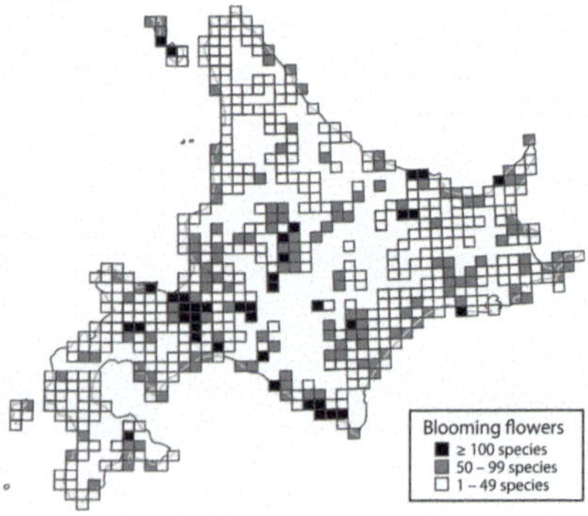


**Fig. 9.7** Hokkaido Flowerthon. **Left**: Volunteers from Tofutsu-ko Waterfowl and Wetland Center, Abashiri; **Right**: Total number of blooming species at each mesh in 2022 (by courtesy of Hokkaido Flowerthon)

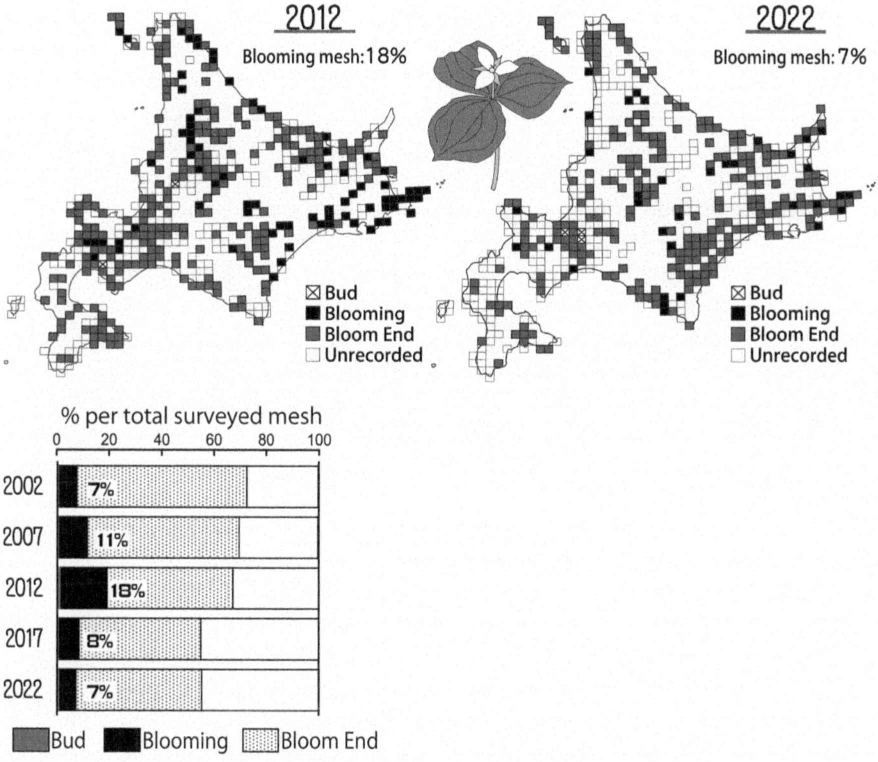


**Fig. 9.8** The blooming map and mesh ratio of *Trillium camschatcense*, a spring ephemeral plant endemic to Hokkaido and northern Honshu, Japan. In 2012, blooming was recorded in 18% of total surveyed meshes, but only in 7% in 2022. Warming trends may be affecting the phenology of these seasonal flowering plants (Watanabe and Niwa 2022: p. 13–14, by courtesy of Hokkaido Flowerthon, partially modified)

and this is also probably a factor in the smooth running of the Flowerthon (Watanabe 2012).

## 9.4 Citizen Science and Museums in the Digital and Internet Age

Needless to say, the development of digital technology since the 1990s has dramatically changed our lives and society. In particular, the digitization of various data and the ability to manage them as databases has brought about a revolution for museums and for citizen science.

Collecting and archiving materials related to natural history is an essential primary function of natural history museums. The digitization of materials and the widespread use of digital photography gave rise to a museum citizen science project in 1994—the Image Database of Fishes in the Kanagawa Prefectural Museum of Natural History, a photograph database of wild fishes. Photos are submitted to the museum, and the ichthyology curator identifies them and adds them to the database. The database is mirrored to online web servers of both the KPMNH and the National Museum of Nature and Science, Tokyo (Fig. 9.9). Many of the contributors are scuba divers who enjoy underwater photography (Matsuura and Senou 2002). As of April 2023, approximately 300,200 pictures (227,700 observations) had been registered, and nearly 90% of them have been uploaded to web servers (H. Senou, pers. comm.) These data provide useful information for compiling local fish faunal lists and monitoring invasive species (Senou et al. 1998; Miyazaki et al. 2015). Although there are other natural history image databases, a significant point here is that the images are archived at a museum (Matsuura and Senou 2002; Miyazaki et al. 2015). This gives the system an academic advantage by ensuring scientific verifiability (Miyazaki 2016).

The internet developed rapidly in the 2000s and its infrastructure has become a major driving force citizen science (Brenton et al. 2018; Wynn 2017). There are many examples of online technologies broadening the base of citizen science participants, such as web and smartphone applications that collect information on occurrence data (Kerstes et al. 2019; Lemmens et al. 2021). In Japan, Biome is a smartphone application for collecting biological occurrence information that incorporates elements of gaming. It has been downloaded 430,000 times and has attracted many participants (Biome 2022). The use of social media such as Twitter (currently X) and Facebook is also increasing (Davis et al. 2017; Kagawa et al. 2020; Pizá et al. 2023). Employing existing social media can keep down initial investment and is expected to have growing potential. In Japan too, NHMs are conducting citizen scientific surveys of this kind. In an example, Takeshi Wada of the Osaka Museum of Natural History collected information on whether barn swallows (*Hirundo rustica*) were nesting in railroad stations in the Kinki region via mailing lists and Twitter in 2012. A total of 124 volunteers cooperated and successfully collected nesting information from all 1536 railroad stations in the target area within about 2 months. This revealed that nesting was influenced negatively by urbanization of land areas and increasing road density but positively by farming and river density (Fig. 9.10; Osawa 2015; Osawa and Wada 2016). In the present author's work, 1304 occurrences of the introduced apple snail (*Pomacea* spp.) were recorded in Japan from 2017 to 2019 by asking people via Twitter and members of the FoM to take photos of the distinctive egg masses and provide their location information (Fig. 9.11; Ishida 2020a). Using social media for a survey will have the effect of making people who are not normally interested in the natural environment aware of the ecology of living creatures around them. In addition, participants can communicate directly with project scientists through the social network. Osawa and Wada (2016) noted that social media can play a significant role in promoting public participation in scientific research.

**Fig. 9.9** Photographs archived in the Image Database of Fishes in the KPMNH are searchable on the web. Screenshot image is the FishPix site at the NSMT (https://fishpix.kahaku.go.jp/fishimage-e/index.html), one of the mirrors of the database (by courtesy of NSMT, KPMNH and Keido Uchino)

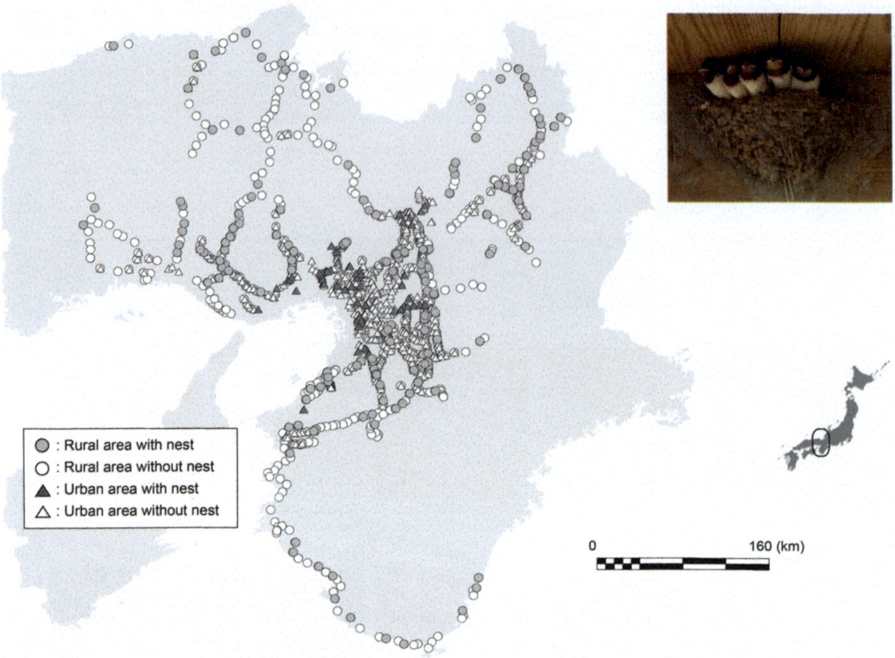


**Fig. 9.10** Locations of railroad stations with and without barn swallow nests, surveyed by volunteers. Rural and urban areas are defined by characterizing the land use within a 5000-m buffer of each station (Osawa 2015: Fig. 2 with permission, partially modified. Swallow's nest and fledglings photo by Akiko Yamanaka)

**Fig. 9.11** Occurrence distribution of the introduced apple snail *Pomacea* spp. reported by volunteers (Drawn from the data in Ishida 2020a)

## 9.5 Challenges and Prospects

I have here reviewed case studies of citizen science in Japan and its relationship with natural history museums over a period of almost 50 years. As we can see, the methods used, the results obtained, and the philosophy of citizen science embody the essence of the regional approach, research, and educational activities of NHMs. This is not limited to Japan, however, and any country with NHMs should have a similar history (Sforzi et al. 2018).

As the above examples indicate, the number of volunteers who participate and cooperate in museum-led citizen science is remarkable. By inference, we believe that this is largely due to the fact that museums are trusted as educational and research institutions. It is only natural that participants would want their data to be used reliably and appropriately. In the case of reporting observation data, however, this means providing activity records that include the observer's own location. In recent years, this has been recognized as sensitive personal information, and some people are naturally cautious about providing it (Ishida 2019). Despite this, many collaborators may be willing to cooperate with NHMs because they believe that they can be trusted to manage the information. In a US survey, Americans said that museums are second only to friends and family in terms of trustworthiness, scoring more highly than individual researchers and/or scientists (American Alliance of Museums 2021). Although no similar survey has yet been published in Japan, this might serve as an insight into the affinity between NHMs and citizen science.

Finally, there are three issues that need to be resolved regarding citizen science in NHMs. The first is the opening of data. Citizen science conducted by NHMs in Japan has been published in the form of articles and publications, but in many cases, the original data is not available or accessible. Open data practices are not sufficiently widespread in the natural sciences, but a standard model is developing through examples such as eBird and iNaturalist, which are seamlessly linked to the Global Biodiversity Information Facility (GBIF), where data from tens of millions of observations can be easily downloaded and used for research and conservation. In the Japanese citizen science field, Monitoring Site 1000, hosted by the Ministry of the Environment, has made almost all its observation data publicly available, and as of 2020, 350,000 records have been provided to GBIF and BISMaL (Biological Information System for Marine Life, managed by the Japan Agency for Marine-Earth Science and Technology) (Ministry of the Environment 2023). Data opening is the simplest and most visible way to give back to society. The use of data papers, which have become popular in recent years, may be another solution. Data papers also have the advantage of being included in academic article databases and thus easy for other scientists to find. For example, the occurrence location data for apple snails from a citizen science project conducted by Ishida (2020a) was subsequently used to validate a model to predict their overwintering areas on the basis of climate data (Yoshida et al. 2022).

The other two issues are funding and fostering the next generation. Kobori et al. (2016) have already cited both as challenges for citizen science in Japan, but I would like to consider them again here.

Funding is important for citizen science, which is not simply a cheap way to collect data by relying on free volunteers. It requires careful management, and human resources are essential to its operation. Natural history museums have staff who are skilled in research and education, but maintaining web systems, as mentioned above, and conducting public relations are naturally expensive (Ishida 2020b). Many museums in Japan suffer from chronic underfunding, and some primary activities rely on external financing. The importance of citizen science is not yet fully recognized by the competitive research funding framework in Japan. In the United States, the American Geophysical Union and other groups proposed in 2022 that the National Science Foundation establish a funding track for community science, and the NSF is considering it (American Geophysical Union 2022). Requests for and implementation of official funding for citizen science are expected to become global trends. We hope that Japan too will engage with this issue.

**Fig. 9.12** **Left**: The Museum of Nature and Human Activities, Hyogo holds "Kyōsei no hiroba" (means symbiosis room), an annual meeting where the museum curators, local naturalist groups and school students present their research and activities since 2006 (photo by Hiromune Mitsuhashi); **Right**: Osaka Natural History Festival (Osaka Shizenshi Festival) has brought together about 100 local nature fan groups and research circles at the Osaka Museum of Natural History for 2 days in November since 2003

We must also engage seriously with the fostering of the next generation. In particular, the aging demography of taxonomists and others skilled in the identification of organisms is a global problem, and training the next generation is an urgent issue (Hopkins and Freckleton 2002; Frobel and Schlumprecht 2016). For NHMs, this is both a challenge and a mission. NHMs have a social education function and practice the fostering of future naturalists on a daily basis. The recruiting and training of the next generation cannot occur without collaboration between universities, researchers, and NGOs (Sforzi et al. 2018), however, and NHMs are expected to be at the core. To cite a relevant example, some NHMs in Japan have organized meetings or festivals as a venue for local nature fan groups and research circles to come together and interact (Fig. 9.12). This may encourage new participants to join in these activities. In addition, the Hokkaido University Museum offers a recurring "parataxonomist" training course to improve citizens' ability to identify organisms (Ohara 2005). Such ongoing efforts are also important.

There are high expectations for NHMs' role and function in conserving biodiversity and achieving a sustainable society (e. g., Ballard et al. 2017). Citizen science is one of the primary mechanisms through which NHMs can advance such practices. If NHMs fully address issues such as those outlined here, their activities should make a significant contribution to resolving them.

## References

American Alliance of Museums (2021) Museums and Trust 2021. https://www.aam-us.org/2021/09/30/museums-and-trust-2021/. Accessed 12 Apr 2023

American Geophysical Union (2022). https://resiliencethruscience.org. Accessed 12 Apr 2023

Ballard HL, Robinson LD, Young AN, Pauly GB, Higgins LM, Johnson RF, Tweddle JC (2017) Contributions to conservation outcomes by natural history museum-led citizen science: examining evidence and next steps. Biol Conserv 208:87–97

Biome (2022) The 3rd anniversary of biome. https://biome.co.jp/3rd_anniversary/. Accessed 12 Apr 2023

Brenton P, von Gavel S, Vogel E, Lecoq M-E (2018) Technology infrastructure for citizen science. In: Hecker S, Haklay M, Bowser A, Makuch Z, Vogel J, Bonn A (eds) Citizen science: innovation in open science, society and policy. UCL Press, London, pp 63–80

Davis A, Major RE, Taylor CE, Martin JM (2017) Novel tracking and reporting methods for studying large birds in urban landscapes. Wildl Biol 2017:1–7

Frobel K, Schlumprecht H (2016) Erosion der Artenkenner. Naturschutz und Landschaftsplanung 48:105–113. (in German)

Fukuda H (1991) Seasonal migration of the Chestnut Tiger, *Parantica sita* in Japan. Insectarium 28(12):4–13. (in Japanese)

Hamada K (2015) Chapter 4.2. Friends of museum and volunteers—diversification of the bearers of museum activities. In: Annual report of the research based on the "General survey of museums" in Japan, FY2014 (JSPS KAKENHI JP25282079). (in Japanese). http://www.museum-census.jp/report2014/report4-2.pdf

Hamaguchi T (1989) Surveys of the distribution of dandelions in Kanagawa Prefecture and its future problems. Natural History Report of Kanagawa (10):9–16. (in Japanese)

Hiura I (1983a) Journey of the chestnut tiger. Nature Study 29(6):7–11. (in Japanese)

Hiura I (1983b) Let's investigate the chestnut tiger in Osaka. Nature Study 29(6):6. (in Japanese)

Hokkaido Flowerthon (2022). https://flowerthon.net. Accessed 16 Apr 2023

Hopkins GW, Freckleton RP (2002) Declines in the numbers of amateur and professional taxonomists: implications for conservation. Anim Conserv 5:245–249

Hotta M (1975a) The distribution of dandelions in Osaka Prefecture—distribution ratio of *Taraxacum japonicum* and *T. officinale* as an environmental indicator. Nature Study 21(4):2–5. (in Japanese)

Hotta M (1975b) The distribution of dandelions in Osaka Prefecture—distribution ratio of *Taraxacum japonicum* and *T. officinale* as an environmental indicator—II. Nature Study 21(5):7–8. (in Japanese)

Hotta M (1977) On the distribution of dandelion (*Taraxacum*) in the Kinki district. Shizenshi-Kenkyu, Occasional Papers from the Osaka Museum of Natural History 1:117–134. (in Japanese with English abstract)

Ishida S (2019) Advantages and problems of citizen science using Twitter—a case study in #Kakichosa. Japanese Council of Science Museums Newsletter 49(6):4–6. (in Japanese)

Ishida S (2020a) Distribution records of apple snails (*Pomacea* spp. ) in Japan collected during 2017–2019 through a citizen science project for introduced species conducted by the Osaka Museum of Natural History. Ecol Res 35(6):1114–1118

Ishida S (2020b) A citizen science survey of the alien land snail *Rumina decollata* in Osaka Prefecture—the methodology and the resulting picture of distribution. Venus 78:105–108. (in Japanese with English abstract)

Japanese Association of Museums (1963) Special edition: Association of Friends of Museums. The Museum Studies 36(7):1–19. (in Japanese)

Japanese Association of Museums (1979) Special edition for the museum friends association. Mus Stud 14(10):3–68. (in Japanese)

Japanese Association of Museums (2009) Report of the comprehensive research study for museums in Japan, FY2008. (in Japanese). https://www.bunka.go.jp/seisaku/bijutsukan_hakubutsukan/shinko/hokoku/h20/1409474.html

Kagawa O, Uchida S, Yamazaki D, Osawa Y, Ito S, Chiba S, The Green-Costumed Snail's Citizen Researchers (2020) Citizen science via social media revealed conditions of symbiosis between a marine gastropod and an epibiotic alga. Sci Rep 10:19647

Kanazawa I, Kaneda S, Fujino M, Yoshimoto T, Watanabe Y (2017) The amazing world of Chestnut Tiger butterfly! Dainippon Tosho, Tokyo, 24 pp. (in Japanese). http://www.dainippon-tosho.co.jp/newsletter/detail.php?id=330

Kano Y (2003) A brief history of research on the long-distance migration of *Parantica sita* in Japan. In: Miyatake Y, Fukuda H, Kanazawa I (eds) Gekkan-Mushi books 6: migrating butterfly, *Parantica sita*. Mushi-sha, Tokyo, pp 57–62. (in Japanese)

Kerstes NAG, Breeschoten T, Kalkman VJ, Schilthuizen M (2019) Snail shell colour evolution in urban heat islands detected via citizen science. Commun Biol 2:264

Kobori H, Dickinson JL, Washitani I, Sakurai R, Amano T, Komatsu N, Kitamura W, Takagawa S, Koyama K, Ogawa T, Miller-Rushing AJ (2016) Citizen science: a new approach to advance ecology, education, and conservation. Ecol Res 31:1–19

Lemmens R, Antoniou V, Hummer P, Potsiou C (2021) Citizen science in the digital world of apps. In: Vohland K et al (eds) The science of citizen science. Springer, Cham. https://doi.org/10.1007/978-3-030-58278-4_23

Makino T 1904. *Taraxacum* in Japan. Shokubutsugaku Zasshi 18:92–93. (in Japanese)

Matsuura K, Senou H (2002) Fish databases in Japan with special reference to fish-image database and its role in biodiversity study. Research report from National Institute for Environmental Studies, Japan (171):220–227. https://www.nies.go.jp/kanko/kenkyu/enpdf/r-171-2002.pdf

Ministry of the Environment (2023) Monitoring site 1000. https://www.biodic.go.jp/moni1000/. Accessed 13 Apr 2023

Miyatake Y, Fukuda H, Kanazawa I (eds) (2003) Gekkan-Mushi books 6: migrating butterfly, *Parantica sita*. Mushi-sha, Tokyo. 241 pp. (in Japanese)

Miyazaki Y (2016) The relationship between citizen science and biodiversity information databases. Jpn J Ecol 66:237–246. (in Japanese)

Miyazaki Y, Murase A, Senou H (2015) A natural history museum as a platform for accumulating verifiable information on non-native fishes: a Japanese example. Manag Biol Invasions 6:105–110

Nakaguchi Y, Masuda H, Nakajo T, Yamanaka K, Satoguchi Y, Analysis Team of Yodo River Research Group (Project Y, Osaka Museum of Natural History) and Mizuhashikake (Lake Biwa Museum) (2013) Survey report of regional distributions of chemical component in the Yodo River system. Bull Osaka Museum Nat Hist 67:45–81. (in Japanese with English abstract)

Ogawa K (1979) Distribution of native and introduced dandelions in the metropolitan area of Japan. Vegetation und Landschaft Japans: Festschrift für Reinhold Tüxen (Bulletin of the Yokohama Phytosociological Society, 16). The Yokohama Phytosociological Society, Yokohama, pp 415–421

Ogawa K, Mototani I (1985) Invasion of the introduced dandelions and survival of the native ones in the Tokyo metropolitan area of Japan. Jpn J Ecol 35:443–452

Ohara M (2005) Parataxonomist training courses by the 21st century COE program "neo-science of natural history" at Hokkaido University. Zool Sci 22:1426

Osawa T (2015) Importance of farmland in urbanized areas as a landscape component for barn swallows (*Hirundo rustica*) nesting on concrete buildings. Environ Manag 55:1160

Osawa T, Wada T (2016) The potential for broad-scale research based on a citizen science approach: a nest-site survey of *Hirundo rustica* at railroad stations in the Kinki District, south-central Japan. Bird Res 12:R1–R8. (in Japanese with English abstract)

Pizá J, Spagnuoli JG, Dop NS, Cazzaniga NJ (2023) From the Mediterranean to Yungas and Patagonia. Dispersal of the non-native gastropod *Rumina decollata* in Argentina. Biologia 78:865–871

Sakuma D (2015) Chapter 4.1. Importance of 'Friends of Museum community' as a basis of cooperation with citizens—relationship to volunteers/regional collaboration. In: Annual report of the research based on the "General survey of museums" in Japan, FY2014 (JSPS KAKENHI JP25282079). (in Japanese). http://www.museum-census.jp/report2014/report4-1.pdf

Senou H, Makiguchi H, Takeya H (1998) List of the fishes of Atami, the eastern coast of the Izu Peninsula, Sagami Bay, on the basis of the underwater photographs registered to KPM-NR. Natural History Report of Kanagawa (19):19–28. (in Japanese with English abstract)

Sforzi A, Tweddle J, Vogel J, Lois G, Wägele W, Lakeman-Fraser P, Makuch Z, Vohland K (2018) Citizen science and the role of natural history museums. In: Hecker S, Haklay M, Bowser A,

Makuch Z, Vogel J, Bonn A (eds) Citizen science: innovation in open science, society and policy. UCL Press, London, pp 429–444

Shiga T, Aquatic Plant Research Team of Yamatogawa River Research Group (Project Y, Osaka Museum of Natural History) (2008) Geographic distributions of *Veronica undulata* and alien species, *V. anagallis-aquatica*, and their hybrids in Yamatogawa River system, central Japan. Bull Osaka Museum Nat Hist 62:65–74. (in Japanese with English abstract)

Shiyake S, Beetle Team of Yamatogawa River Research Group (Project Y, Osaka Museum of Natural History) (2008) Fauna and distribution of the riffle beetles (Coleoptera: Elmidae) in Yamatogawa River system, central Japan. Bull Osaka Museum Nat Hist 62:47–64. (in Japanese with English abstract)

Shiyake S, Yasui M, Ichikawa A, Katsura K, Kawai M, Nakatani K, Yamazaki K, Beetle Team of Urban Nature Research Group (Project U, Osaka Museum of Natural History) (2020) Beetle fauna in Osaka City and its transition. Shizenshi-Kenkyu, Occasional Papers from the Osaka Museum of Natural History 4(3):41–104. (in Japanese with English abstract)

Silvertown J (2009) A new dawn for citizen science. Trends Ecol Evol 24:467–471

Suzuki M, Kameda K, Sakuma D, Manabe T (2010) The significance of regional museums' work on natural history (Museum and Ecology (14)). Jpn J Ecol 60:399–403. (in Japanese)

Takibata M (2003) The history of citizens' participation at Osaka Museum of Natural History (2). J Museol Soc Japan 28(2):1–22. (in Japanese with English abstract)

The Executive Committee of Dandelion Survey in West Japan (2016) Report of Dandelion survey in West Japan 2015. The Executive Committee of Dandelion Survey in West Japan, Osaka. (in Japanese). http://gonhana.sakura.ne.jp/tanpopo2015/doc/report2015.pdf

The Nature Conservation Society of Hyogo Prefecture (1971) Request for cooperation in the survey of environmental disruption. Hyogo-ken no Shizen (1):15. (in Japanese)

The Nature Conservation Society of Hyogo Prefecture (1972) Report of the 1st survey of environmental disruption—serious disruption in Hanshin district. Hyogo-ken no Shizen (2):2–14. (in Japanese)

Watanabe O (2012) The possibility of a network for local museums in Hokkaido (Museum and Ecology (18)). Jpn J Ecol 62:89–93. (in Japanese)

Watanabe O, Niwa S (2022) Results and analysis. In: Hokkaido Flowerthon Secretariat (ed) Let's map flowers!—report of the Hokkaido Flowerthon 2022. PIIF Hokkaido Shimbun Press Wildlife Fund, Sapporo, pp 6–18. (in Japanese)

Wynn J (2017) Citizen science in The digital age: rhetoric, science, and public engagement. The University of Alabama Press, Tuscaloosa, 207 pp

Yoshida K, Wada T, Matsukura K, Shiba T (2022) Potential overwintering areas of the alien apple snail, *Pomacea canaliculata*, in Japan at its northern distribution limit. Aquat Invasions 17:402–414

# Part III
# Considering Data and Environments with Participants

# Chapter 10
# Science Communication: Bridging Between Science and Society

**Yuko Ikkatai**

**Abstract** As science and technology have become integral parts of our lives, science communication has become extremely important in bridging the gap between science and society. Science communication includes a variety of activities: science festivals, science cafés, scientific educational programs, science films and TV programs, websites, social media, and researcher outreach efforts. Recently, citizen science has also been considered one of the approaches to science communication. Science communication has developed mainly in Europe. In Japan, it has been promoted since the 1990s, focusing on dialogue with the public under the Basic Plan for Science and Technology. This chapter intends to provide basic knowledge of science communication and the current situation in Japan.

**Keywords** Science · Society · Communication · Model · Practice

## 10.1 What is Science Communication?

While science and technology have become indispensable in our lives, science communication has become important for us to understand and participate in science. Science communication is a concept that encompasses a variety of activities. Burns et al. (2003) defined science communication as "the use of appropriate skills, media, activities, and dialogue to produce one or more of the following five responses (awareness, enjoyment, or other affective responses, interest, opinions, and understanding) to science." First, awareness refers to increasing people's recognition and comprehension of new aspects of science. Second, enjoyment refers to other affective responses that entail finding entertainment or artistry in science and establishing an emotional connection with it. Third, interest refers to individuals spontaneously

Y. Ikkatai (✉)
Institute of Human and Social Sciences, Kanazawa University,
Kanazawa-shi, Ishikawa, Japan
e-mail: y.ikkatai@staff.kanazawa-u.ac.jp


Y. Suzuki-Ohno (ed.), *Community Science in Ecology*, Ecological Research Monographs, https://doi.org/10.1007/978-981-97-0304-3_10

developing an inclination toward science and engaging in activities or communication related to science. Fourth, opinions involve shaping, reshaping, or reaffirming attitudes and viewpoints related to science. Fifth, understanding aims to comprehend the content of science, the processes within science, and the societal factors affecting science (Burns et al. 2003).

Science communication began with the awareness that it is essential to convey science to the public due to issues such as a lack of scientific literacy and disinterest in science. In 1985, the Royal Society issued a report entitled "The Public Understanding of Science," which focused on the public's comprehension of science. This report emphasized that improving the general level of public understating of science (PUS) is an urgent task for the well-being of the country (the United Kingdom). It provided specific guidance to individuals in various communities, including formal education, mass media, the scientific community, industry, museums, libraries, and more, outlining what each could do to contribute to this goal (The Royal Society of London 1985). This led to the emergence of various science communication initiatives designed to address a perceived deficiency in scientific knowledge among the public. At the time, there was a prevalent belief that the general public's lack of trust in science and technology was primarily attributed to its lack of knowledge of science. In addition, there was a belief that the more experts in science conveyed scientific knowledge to the citizens, the likelier the citizens would understand and develop an affinity for science and technology. This approach is commonly referred to as the deficit model and is criticized (Wynne 1991, 1992, 1993). The deficit model is characterized by the one-way transmission of scientific information from experts to the public, with the expectation that the public needs to "fill the deficit" in their scientific knowledge. However, it has gradually become apparent that the deficit model has limitations. Through debates surrounding issues such as bovine spongiform encephalopathy (BSE), it has become clear that simply acquiring the correct knowledge does not necessarily alleviate public concerns and resistance. In 2000, the House of Lords Select Committee on Science and Technology in the United Kingdom published a report entitled "Science and Technology—Third Report" and emphasized the importance of dialogue between experts and the public (House of Lords 2000). As a result, there has been a shift from emphasizing activities aimed at increasing understanding of science (deficit model) to prioritizing dialogue activities that involve learning from citizens' questions, concerns, and opinions. However, we need to be careful that this is not to deny the one-way exchange of information. Information sharing is an important prerequisite and foundation for scientific communication.

Trench (2008) proposed an analytical framework for a scientific communication model consisting of three models (one-way deficit, two-way dialogue, and three-way participation). Later, Bucchi and Trench (2021) modified this framework into a dissemination, dialogue, and participation model. The authors explained that the three models are dynamically related. The dissemination model is characterized as purposive and closed, while the participation model is more nonpurposive and open. The dissemination model is used for findings as completed knowledge, the dialogue model is used for issues as applications and implications of knowledge, and the participation model is used for processes as interpretations and constructions of knowledge.

Additionally, Bucchi (2013) noted that the evolution of the internet has transformed Science Communication from 1.0 to 2.0. In Science Communication 1.0, the main actors were mediators, journalists, professional communicators, dissemination channels, and science museums. Communication with the public was mediated and filtered. The source of the information and the reputation of the brand were important to determine whether the content was trustworthy (Bucchi 2017). However, in Science Communication 2.0, with the widespread use of the internet, the main actors have changed to research institutions, scientists, and digital media companies, and communication is no longer mediated but transmitted directly to the public (Bucchi 2013). In other words, the public now has access to the process by which science is produced and what is being discussed among experts. Thus, in Science Communication 2.0, researchers and research institutions have a greater responsibility for science communication.

## 10.2 Practicing Science Communication

Science communication plays a crucial role in bridging the gap between the scientific community and the public. These diverse communication practices, such as science festivals, science cafés, scientific educational programs, science films and TV programs, websites, social media, and researcher outreach efforts, aim to make science more accessible and engaging for people of all backgrounds. Some topics, especially those involving complex scientific or technological aspects, require more in-depth discussions and informed citizen engagement. For example, consensus conferences have been developed to address these intricate subjects. A consensus conference started in 1985 in Denmark, and it has since been held worldwide. This is a deliberative process in which a diverse group of citizens, often selected to represent various segments of the population, come together to discuss complex scientific or technological topics. These topics can range from environmental issues to emerging technologies. Past conferences have covered the topics, for example, of air pollution and genetically modified (GM) foods. The primary objective of a consensus conference is to facilitate public engagement and generate a collective recommendation or consensus on the issue being discussed. Through informed discussions, participants work toward a shared understanding and potential solutions that can inform policymakers and contribute to more informed and inclusive decision-making processes. This approach is valuable for addressing contentious or multifaceted issues where public input is essential. Similarly, Worldwide Views, initiated in Denmark in 2009, employs a deliberative process. This approach involves organizing structured dialogues and meetings in various countries simultaneously, allowing citizens to express their opinions, concerns, and preferences on topics such as climate change and biodiversity. Additionally, citizen science is also a science communication approach. Lewenstein (2016) described citizen science as one of the most significant advancements in science communication.

Encouraging the participation of individuals with various backgrounds is a significant challenge in science communication. A Japanese study indicated that the majority of participants in science cafés, public events where scientists discuss and explain scientific topics in a relaxed setting, often in a cafe or similar venue, are those in high interest groups (Kano et al. 2013). Encouraging the participation of individuals with low interest in science can involve various strategies. One approach is to integrate science with seemingly unrelated fields, such as art (Kano et al. 2013; Lesen et al. 2016), or connect it to everyday life themes. Hosting events in commercial venues, such as shopping malls, can increase the chances of engaging people with a low interest in science (Goto and Kano 2021). Additionally, emphasizing storytelling and conveying information with compelling narratives about science can capture their interest (Joubert et al. 2019). Introducing science communication as part of school education or curricula may also help cultivate interest in science among younger generations.

## 10.3 Science Communication in Japan

Currently, science communication is being promoted in Japan at various institutions and levels. For example, the National Museum of Nature and Science defines science communication as the process of connecting science and the general public and establishing science as a culture in society (National Museum of Nature and Science 2017). At the government level, the Japanese Ministry of Education, Culture, Sports, Science, and Technology (MEXT) defines science communication as activities aimed at communicating the fun of science and issues surrounding science and technology to people, thinking together, and raising awareness (Ministry of Education, Culture, Sports, Science and Technology n.d.).

Watanabe (2012) described that 1960 serves as a significant starting point for the spread of science education in Japan. In 1960, Science and Technology Week was established, during which various institutions held events related to science and technology. There was an emphasis on fostering dreams related to science and technology until the 1970s. In 1985, the Tsukuba Science Expo was held, suggesting that there was a focus on science promotion in the 1980s. In the 1990s, efforts were directed toward combating science detachment (Watanabe 2012).

The policy trends in science communication have developed primarily through the Science and Technology Basic Plan. The First Basic Plan for Science and Technology was formulated in 1996. This report highlighted the importance of promoting understanding and generating interest in science and technology. It emphasized the need for researchers to communicate information to society. The Basic Plan for Science and Technology has been updated every 5 years. The Third Basic Plan for Science and Technology, released in 2006, emphasized the development of professionals, including science and technology communicators, to meet the needs of society (Japan Science and Technology Agency n.d.). In 2005, training programs for science communicators were initiated at Hokkaido University, the University of Tokyo, and Waseda University. This year also marked the beginning of an increase in science cafés. Therefore, 2005 is "Year One" of science communication in Japan (Kobayashi 2010).

In 2011, the Great East Japan Earthquake occurred, leading to a significant decrease in trust in scientists immediately after the disaster. When asked if they think scientists are reliable, about 80% of the respondents in the October–November 2010 survey before the earthquake said they were reliable or somewhat reliable. However, in the post-disaster period of May–June 2011, the percentage dropped to about 60% (NISTEP 2012). Science communications were also limited amid the societal turmoil at that time, and criticism arose that science communication was not useful. A study reported that about 50% of science communicators felt the limitations of their own science communication activities after the earthquake, and 80% of science communicators felt that the criticism was valid. They grappled with dilemmas arising from their limited ability to engage due to barriers related to their own skills, expertise, and public emotions (Ikkatai and Yokoyama 2016). The Fourth Basic Plan for Science and Technology, released in 2011, included provisions for promoting risk communication activities during emergencies and emphasized that science communication activities include the training of science communicators and engaging in two-way dialogues with the public (Japan Science and Technology Agency n.d.). The Fifth Basic Plan for Science and Technology in 2016 emphasized co-creation with diverse stakeholders (Ministry of Education, Culture, Sports, Science and Technology (MEXT) 2016), and the subsequent Sixth Basic Plan for Science and Technology and Innovation in 2021 incorporated the promotion of citizen science (Ministry of Education, Culture, Sports, Science and Technology (MEXT) 2021).

## 10.4 Researchers Toward Science Communication in Japan

In a 2013 survey targeting 8964 researchers, the governmental policy of promoting dialogue with the public in the Fourth Science and Technology Basic Plan received support from 70.9% of researchers, suggesting their acknowledgment of the necessity of science communication by the researchers themselves. While about 64.4% of respondents stated that they have been involved in science communication activities, only 28.1% indicated that it has become an ingrained part of their voluntary activities as researchers. In the same survey, there were responses indicating several barriers for researchers to engage in science communication. These barriers include a "lack of time," "the extensive administrative work required for activities," "lack of recognition as part of one's professional achievements," "difficulty in securing funding," and "challenges in establishing venues for communication activities" (Japan Science and Technology Agency (JST) 2013, Koizumi et al. 2013). Furthermore, Mizumachi et al. (2011) identified five factors leading to early-career researchers' reluctance to participate in science cafés: the effort being inconvenient or time-consuming, feeling pressured to represent science appropriately, not aligning with their work scope, not seeing any personal benefits, and having reservations about engaging in public dialogue. This indicates that various forms of support are necessary to establish voluntary science communication activities by researchers in society.

## 10.5 Public Attitudes Toward Science in Japan

A continuous survey conducted from 2009 to 2019 revealed that the level of interest in science and technology among people living in Japan fluctuated between roughly 60% and 70%. Additionally, when considering sex differences, the trend that men tend to have a higher level of interest in science and technology compared to women is relatively consistent (Hosotsubo et al. 2020). The Victorian segmentation method, which segments people based on their "interest in science and technology" developed by the government of Victoria, Australia, revealed that the proportions of the interest group, potential interest group, and low-interest group in 2013 were 16.1%, 61.4%, and 22.6%, respectively (PESTI 2016; Kano 2014). In other words, the largest proportion of people had a moderate level of interest in science. This trend was also found in a 2020 survey that targeted people in Japan ($n = 1108$) and the United States ($n = 1063$). The interest group occupied the largest (about 60%) in the United States, while in Japan, the potential interest group was the largest (about 40%) (Hartwig et al. 2023). Another study in 2016 investigated public attitudes toward participating in dialogue and collaborative initiatives, such as consensus conferences and Worldwide Views. Analysis of data collected from 1000 people in Japan revealed that approximately 30% of respondents expressed their interest in participating in such initiatives in the future. When asked about the stage at which they believed dialogue and collaboration to be necessary, around 40% of respondents indicated that dialogue was required either before or shortly after the commencement of research and development (Japan Science and Technology Agency (JST) 2016).

On the other hand, the level of trust in scientists is not very high in Japan. In a global survey conducted from 2019 to 2020, only 23% of respondents in Japan answered that they have "a lot" of trust in scientists. This percentage was lower than in India (59%), Germany (43%), the United Kingdom (42%), and the United States (38%) (Pew Research Center 2020). These findings indicate the need for continued efforts in science communication and public engagement to bridge these gaps in interest and trust.

## 10.6 Conclusions

In this chapter, the author reviewed the beginnings and various models of science communication. The author also explained how science communication was introduced in Japan and the level of interest and trust in science among the Japanese people. The importance of science communication is gradually being recognized in Japan. Human resources are also being developed. In the future, it is necessary to further increase understanding and interest in science through the development of more effective science communication strategies tailored to the Japanese context.

**Acknowledgments** This work was supported by JSPS KAKENHI Grant Number 20K14114.

## References

Bucchi M (2013) Style in science communication. Public Underst Sci 22(8):904–915. https://doi.org/10.1177/0963662513498202

Bucchi M (2017) Credibility, expertise and the challenges of science communication 2.0. Public Underst Sci 26(8):890–893. https://doi.org/10.1177/0963662517733368

Bucchi M, Trench B (2021) Rethinking science communication as the social conversation around science. J Sci Commun 20:Y01. https://doi.org/10.22323/2.20030401

Burns TW, O'Connor DJ, Stocklmayer SM (2003) Science communication: a contemporary definition. Public Underst Sci 12:183–202. https://doi.org/10.1177/09636625030122004

Goto T, Kano K (2021) Evaluating the representation of participants in a scientific workshop held in a shopping mall: a challenge for attracting a wide range of participants in terms of interests and learning motivation in science (in Japanese). Japan J Educ Technol 45:113–126. https://doi.org/10.15077/jjet.44059

Hartwig T, Ikkatai Y, Takanashi N, Yokoyama MH (2023) Artificial intelligence ELSI score for science and technology: a comparison between Japan and the US. AI Soc 38:1609–1626. https://doi.org/10.1007/s00146-021-01323-9

Hosotsubo M, Tsunoda H, Kano K, Okamura A, Hoshino T (2020) Public attitudes to science and technology: social acceptance of new technologies (in Japanese). National Institute of Science and Technology Policy. https://doi.org/10.15108/rm296

House of Lords (2000) Science and society—third report. https://publications.parliament.uk/pa/ld199900/ldselect/ldsctech/38/3801.htm. Accessed 14 Sept 2023

Ikkatai Y, Yokoyama H (2016) What could science communicators do after the great East Japan earthquake? (in Japanese). Jpn J Sci Commun 19:57–70. https://doi.org/10.14943/74100

Japan Science and Technology Agency (n.d.) Deepening the relationship between science and society (in Japanese). https://www.jst.go.jp/sis/scienceinsociety/. Accessed 14 Sept 2023

Japan Science and Technology Agency (JST) (2013) Report on the Questionnaire Survey on Science Communication Activities by Researchers (in Japanese). https://www.jst.go.jp/sis/archive/items/csc_fy2013_03.pdf. Accessed 14 Sept 2023

Japan Science and Technology Agency (JST) (2016) Survey report for fiscal year 2015: facing social expectations and concerns of science and technology—Challenges in practicing dialogue and collaboration (in Japanese). https://www.jst.go.jp/sis/scienceinsociety/investigation/items/report2015_01.pdf. Accessed 14 Sept 2023

Joubert M, Davis L, Metcalfe J (2019) Storytelling: the soul of science communication. J Sci Commun 18: E. https://doi.org/10.22323/2.18050501

Kano K (2014) Toward achieving broad public engagement with science, technology, and innovation policies: trials in Japan vision 2020. J Deliberative Mech Sci 3:1–23. https://doi.org/10.4471/demesci.2014.12
Kano K, Mizumachi E, Iwasaki T, Isobe H, Kawahito Y, Maenami H (2013) Segmentation and targeting of participants in science cafes: from the viewpoint of the extent of engagement in science and technology (in Japanese). Jpn J Sci Commun 13:3–16. https://doi.org/10.14943/61989
Kobayashi T (2010) Scientific knowledge in society and it's communication (in Japanese). Kagaku tetsugaku 43(2):33–45. https://doi.org/10.4216/jpssj.43.2_33
Koizumi A, Morita Y, Kawamoto S (2013) Reward research outreach in Japan. Nature 500(7460):29–29. https://doi.org/10.1038/500029a
Lesen AE, Rogan A, Blum MJ (2016) Science communication through art: objectives, challenges, and outcomes. Trends Ecol Evol 31:657–660. https://doi.org/10.1016/j.tree.2016.06.004
Lewenstein B (2016) Can we understand citizen science? J Sci Commun 15: E. https://doi.org/10.22323/2.15010501
Ministry of Education, Culture, Sports, Science and Technology (n.d.). What is science communication? (in Japanese). https://www.mext.go.jp/kids/find/kagaku/mext_0005.html. Accessed 14 Sept 2023
Ministry of Education, Culture, Sports, Science and Technology (MEXT) (2016) The 5th science and technology basic plan (in Japanese). https://www8.cao.go.jp/cstp/kihonkeikaku/5honbun.pdf. Accessed 14 Sept 2023
Ministry of Education, Culture, Sports, Science and Technology (MEXT) (2021). The 6th science, technology and innovation basic plan (in Japanese). https://www8.cao.go.jp/cstp/kihonkeikaku/6honbun.pdf. Accessed 14 Sept 2023
Mizumachi E, Matsuda K, Kano K, Kawakami M, Kato K (2011) Scientists' attitudes toward a dialogue with the public: a study using "science cafes". J Sci Commun 10(4):A02. https://doi.org/10.22323/2.10040202
National Museum of Nature and Science (ed) (2017) How to start science communication (in Japanese). Maruzen Pubilishing
NISTEP (2012) Survey on changes in public attitudes toward science and technology-from the results of the monthly internet-based attitude survey and interview surveys (in Iapanese). https://www.mext.go.jp/component/b_menu/shingi/toushin/__icsFiles/afieldfile/2013/03/15/1331441_03.pdf. Accessed 14 Sept 2023
PESTI (2016) Public opinion survey on science, technology and innovation policy publication of report and individual data on "Survey on Attitudes toward Life and Science and Technology" (in Japanese). https://www.nistep.go.jp/research/scisip/data-and-information-infrastructure/pesti-data. Accessed 14 Sept 2023
Pew Research Center (2020) Science and scientists held in high esteem across global publics. https://www.pewresearch.org/science/2020/09/29/science-and-scientists-held-in-high-esteem-across-global-publics/. Accessed 14 Sept 2023
Royal Society of London (1985) The public understanding of science. The Royal Society of London
Trench B (2008) Towards an analytical framework of science communication models. In: Cheng D, Claessens M, Gascoigne T, Metcalfe J, Schiele B, Shi S (eds) Communicating science in social contexts: new models, new practices. Springer, pp 119–135. https://doi.org/10.1007/978-1-4020-8598-7_7
Watanabe (2012) Toward science communication 2.0 (in Japanese). J JASC 1:6–11
Wynne B (1991) Knowledges in context, science. Technol Human Values 16:111–121. https://doi.org/10.1177/016224399101600l0
Wynne B (1993) Public uptake of science: a case for institutional reflexivity. Public Underst Sci 2:321–337. https://doi.org/10.1088/0963-6625/2/4/003
Wynne B (1992) Misunderstood misunderstanding: social identities and public uptake of science. Public Underst Sci 1:281–304. https://doi.org/10.1088/0963-6625/1/3/0

# Chapter 11
# The Essential Principle for Appropriate Data Policy of Citizen Science Projects: Correspondence for the Diverse and Transmutable Attributes of Participants

**Takeshi Osawa**

**Abstract** *Citizen science* is one of new paradigms of science. This concept features various project forms, participants, and motivations and implies the need for attention to ethical issues for every participant, which frequently includes nonacademics. In this chapter, I address ethical issues associated with citizen science projects that focus on the data treatment rule and demonstrate a concept on appropriate data policy for these projects. First, I demonstrate that citizen science projects tend to include different types of collaboration, which may lead to certain conflicts among participants in terms of data sharing. Second, I propose an idea that could integrate different types of collaboration according to the theory *transcend*. Third, I take a case of a citizen science project through which transcend occurred and elucidate the difference between ordinal research and citizen science projects, specifically in terms of the goals of these projects and the goals and motivations of participants, which may change. Finally, I proposed one conceptual idea on how the principal investigator (PI) of a citizen science project can establish data policy after assessing the rights of participants. The basic idea is the division and organization of the data policy in a hierarchy for the project and for the participants. Data policy is one of the important items for establishing the appropriate methods for citizen science as new style of science. As such, practice and framing related to data policy must be carefully monitored and reflected on.

**Keywords** Collaboration type · Data sharing · Ethical issue · Open science · Transcend

---

T. Osawa (✉)
Graduate School of Urban environmental Sciences, Tokyo Metropolitan University, Hachiouji, Tokyo, Japan



Y. Suzuki-Ohno (ed.), *Community Science in Ecology*, Ecological Research Monographs, https://doi.org/10.1007/978-981-97-0304-3_11

## 11.1 Introduction

Public participation in scientific projects is expanding globally, which is labeled as *citizen science* (Heigl et al. 2019) or other similar terms such as "citizen-driven projects," "participant-led research," and "community-based participatory research" (Vayena and Tasioulas 2013; Kondo et al. 2019). However, several forms of project exist within this single concept. For example, each project pursues different goals, such as enjoyment, education, and publication (Kobori et al. 2016; Sasaki et al. 2016; Osawa 2019). In addition, projects recruit various types of participants in terms of number and motivation (Sasaki et al. 2016; Chap. 7, Editor's Note). These structural differences compare with traditional science could cause regulatory gaps and mismatches among participants (Guerrini et al. 2018).

One of the crucial gaps and mismatches in a citizen science project is asymmetries in ethical points among participants (Kondo et al. 2019). For example, a number of citizen scientists may assert ownership of the contributed data and desire to control data usage, but PIs in academics frequently fail to consider these aspect at the beginning of a project (Mahr and Dickel 2019). For example, a few participants may aim for scientific publication (i.e., a project goal), such that they press a contribution that for all participants; however, this pressure may be a nuisance for participants who are not interested in this objective (Sasaki et al. 2016). In the Internet era, citizen science projects could involve millions of participants (G7 2019), but such an expansion fomented the emergence of such asymmetries.

Hayden (2012) indicated that problems, such as an inadequate process of providing informed consent, occurred in many citizen science projects (Hayden 2012). Unfortunately, scholars point out that this process is vague for several citizen science projects (Vayena and Tasioulas 2013). A typical case is the use of the crowdsourcing approach, which could potentially recruit millions of people from all over the world through online platforms (Welvaert and Caley 2016; Picaut et al. 2019). Several PIs of this type of citizen science project may fail to provide participants with appropriate information; in addition, several participants may fail to carefully check the materials for informed consent, such as documents that explain intellectual property rights, in relation to the project. However, additional strict and complex procedures may not provide participants with benefits but lead to confusion and risk instead, which could deter potential participants from these projects (Hayden 2012).

At the global level, citizen science projects largely contribute to science development and provide benefits for contributors and added value for society (Silvertown 2009; Dickinson et al. 2010; Kobori et al. 2016; Vohland et al. 2019). Alternatively, many researchers highlight the necessity of the ethical discussion of citizen science projects (Vayena and Tasioulas 2013; Kondo et al. 2019; Rasmussen and Cooper 2019; Lee et al. 2020). The significant ethical challenges of these projects are frequently related to data issues, which include exploitation, ownership, and sharing (i.e., data policy; Rasmussen and Cooper 2019; Lee et al. 2020). Therefore, in this chapter, I discuss ethical issues associated with citizen science projects and especially focus on the treatment of collected data and the project management of PIs with regard to this issue.

## 11.2 The Basic Principle of Data Sharing and Citizen Science Projects

Although labeled as citizen science, it is only one of many research strategies that is, at least, specific to academics. Thus, data policy should essentially be the same with that used in ordinal research collaboration in academics. Prior to beginning an ordinal research project, PIs decide on data ownership and the rights of the participants. Recently, funding agencies request projects to submit a data management plan (Gajbe et al. 2021). As such, PIs of citizen science projects should establish their data policy.

The simplest rule of data policy is "data owned by the collector." When an individual wants to integrate all data sets in a project and use it to fulfill the research objectives, then each data owner should give permission, even if he/she is one of participant of the project. Another simple method for data policy refers to "data evenly owned by all participants." Under this policy, all participants can use data without specific permissions within the project. If an outsider to the project desires to use the data set, then they should obtain permission from the member(s) of the citizen science project. To provide permission, consensus within the project may be required.

However, several projects employ top-down approaches, such as decision-making, to use data owned by the PI of the project. To avoid this scenario, Buyx et al. (2017) demonstrate the concept "every participant is a PI." This idea requires the involvement of participants in all tasks, such as all executive decisions, including the setting and planning of the research agenda planning and data management and use, among others, in addition to data collection.

These rules could ensure that all participants hold the same rights for the data policy, that is, a fair one. These rules are simple and acceptable for many participants who want to use the data for their research. However, this principle is not a *one-size-fits-all* provision for all cases of citizen science projects, because the data set is only one of the deliverables instead of the overall achievement of the project. Citizen science projects, in which nonacademics frequently participate, lack interest in data usage. A few of them may not want to use the data, because their motive for the project is not academic publication but enjoyment and educational values, among others. In addition, many participants may prefer not to hold authority for the project in reality. As such, this over-duty could become a burden for participants. In this situation, participants who are uninterested in using the data can decline the right to data use, which is the simple answer. However, accepting this rule is frequently difficult for participants, because a few of them may wish to provide data with specific persons instead of all participants.

## 11.3 Different Types of Collaboration and Their Coexistence

Why then does such a complex situation occur in citizen science projects? One possible explanation is that one citizen science project not only frequently recruits several types of participants but also includes different styles of collaboration among participants. In this study, I demonstrate two typical styles of collaboration in a citizen science project according to Matsubara and Okoso (2022); Fig. 11.1. The first type of collaboration is one in which the participants share the same goal (Fig. 11.1a). In this type of collaboration, the participants hold the same interests and share the goal of the project; thus, they can directly collaborate. This type is a general case ordinal collaboration in academics, because all participants basically aim to publish their project achievements; thus, they exert a united effort for this common goal.

The second type is indirect or understudy collaboration in which one or a few members have their interests and goal and others support this goal, or both have a common interest or goal, but one or some representative participants achieve with others supports (Fig. 11.1b). In this collaboration style, the participants could work toward achieving the goal with consent. This type may be used in many citizen science projects such as publication that uses the crowdsourcing approach (Osawa et al. 2017; Njue et al. 2019). Specifically, PIs coordinate the collection of data and use them for publication, whereas participants contribute to data collection only.

**Fig. 11.1** Two typical types of collaboration among participants

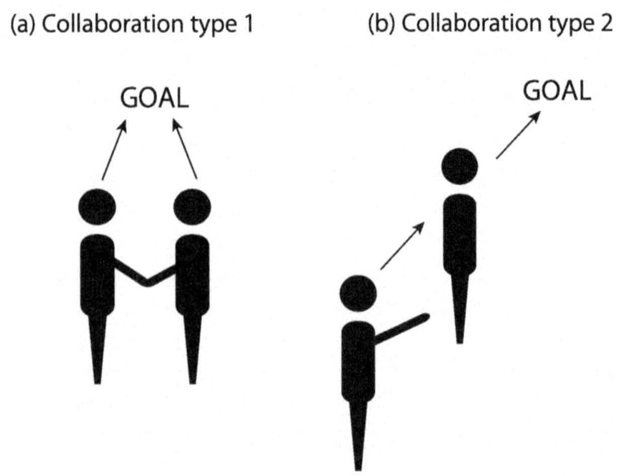

If one project includes one of the two types of collaboration, then all participants could satisfy their motivation. However, such situation is a rare in recent citizen science projects, which frequently employ large numbers of participants. In general, even for one citizen science project, both types of collaboration could be simultaneously used. For example, participants want to use data (i.e., type-1 collaboration), but other want to provide data for specific persons (i.e., type-2 collaboration), in which case a conflict between them could occur. Therefore, PIs should solve the conflict between the types to appropriately treat the data.

How can such different types of collaboration be integrated within one citizen science project? One effective solution is the *appropriate shift*. Kondo et al. (2019) proposed the idea of discovering and sharing goals that actors with different interests can tackle together (Kondo et al. 2019). Matsubara and Okoso (2022) proposed a similar idea. The current study demonstrated one of the realized models on appropriate shift between different types of collaboration (Fig. 11.2) as type-3 collaboration. This type is a common-mode gain in that participants have different interests and goals, but their intermediate goal is common; thus, they can co-benefit to achieve this intermediate goal (Fig. 11.2a). In the type-3 collaboration, types 1 and 2 may co-exist by undertaking the goals of each participant (Fig. 11.2b). For type-3 collaboration, the intermediate goal for one participant could be the final goal for others (Fig. 11.2b). Such types of multiple pathway should be allowed in this type of collaboration (Kondo et al. 2019). If PIs of citizen science projects can coordinate the type-3 collaboration within a project, then conflict due to data treatment could be avoided for a significant portion of the project.

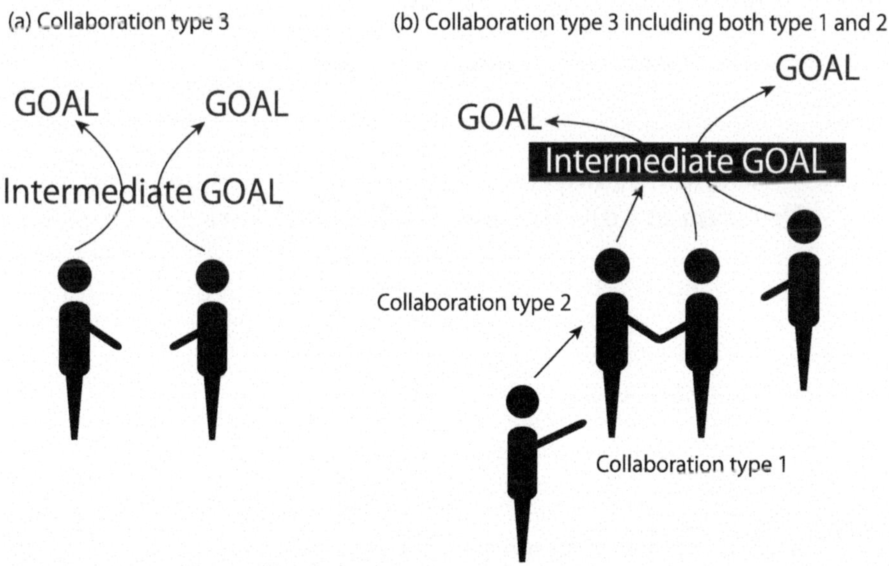


**Fig. 11.2** (**a**) Type-3 collaboration in which participants share a common intermediate goal or different goals. (**b**) This type of collaboration includes types 1 and 2

## 11.4 Appropriate Shift, Boundary Spanning, and *Transcend* Method

The appropriate shift in specifically determining an intermediate goal among participants in a citizen science project could be considered as one of the methods for boundary spanning, which is defined as actions for establishing linkages and managing interactions with parties in an external environment. In the context of citizen science projects, the external environment is a different type of collaboration. Boundary spanning could assist a team and others linked to the team in meeting performance goals and objectives (Marrone 2010). Citizen science projects frequently recruit not only several types of participants but also different types of collaboration style; thus, the boundary spanning approach is required for integrating these elements.

Kondo et al. (2019) proposed that the appropriate shift is one of the *transcend* method for spanning boundaries among the interests of participants. This method fundamentally intends to transform the characteristics of conflict and is an attempt to construct new realities among conflicting parties and to provide a space for the transformation of these contradictions within the conflict. To realize this notion, finding a common, intermediate goal is one of the key items (Kondo et al. 2019). In the transcend method, the presence of a mediator is a prerequisite (Kondo et al. 2019). The role of the mediator is to assist in diverting a conflict point and to facilitate dialogue among participants (Kondo et al. 2019). Once a dialogue emerges and develops, the participants can then communicate and may formulate multiple methods (i.e., intermediate goals). To maintain the common intermediate goal, participants could satisfy their desire despite their diverse goals. The PI of the project or non-PI closely related to the PI can assume the role of a mediator.

## 11.5 Case of the Transcend Method in a Plant Monitoring Program in Fuji–Hakone–Izu National Park

In this section, I discuss a case of transcend in a citizen science project, that is, a monitoring program for flowering plants in Fuji–Hakone–Izu national park in Japan. This program is conducted by volunteers at the park, which is located in the Hakone region (Osawa and Inohara 2008; Osawa 2013).

The Fuji–Hakone–Izu National Park is one of the oldest national parks in Japan. The Hakone region of the national park is located in the town of Hakone, which is one of the biodiversity hotspots in Kanagawa Prefecture (Tanaka 2005). The park volunteer association in the Hakone region has approximately 100 members who have conducted the monitoring of plant species since 2001 along famous hiking trails (Osawa and Inohara 2008; Osawa 2013). The primary objective of monitoring is to deliver information about flowers to visitors; these results are displayed at the Hakone visitor center (Fig. 11.3). The secondary objective is to collect basic

**Fig. 11.3** Latest results on the plant monitoring program at the Hakone visitor center. The trails mainly display flowering plants for visitors

information on biodiversity in the region. For example, if alien invasive plants were newly recorded, then this information can be quickly shared with park managers. Additionally, a few of the participants joined this activity to learn about plant names and take a photo for enjoyment. Thus, this activity highlights the different goals and motivations of participants. Importantly, monitoring records were closed, because they were not appropriately managed at the time (Fig. 11.4a).

Although this monitoring program is not a scientific project, one park volunteer cited the significance of the monitoring records, which he organized and maintained for a few years. However, the other participants and park managers did not agree

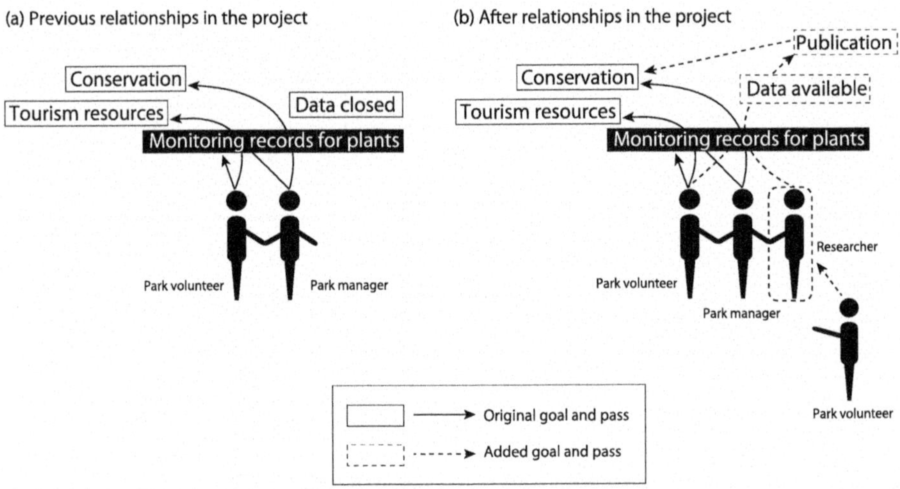


**Fig. 11.4** Diagram of transcend at the plant monitoring program in Hakone national park. The solid line indicates original goals and passes on the project. The dash line indicates the added goals and passes of the participants after added the Researcher

with this significance; thus, he exerted personal effort. Fortunately, I found the records and explained this significance for stakeholders, including monitoring participants, could obtain consent from park volunteer association. I analyzed the monitoring data with the park volunteer and demonstrated the important trend on the drivers of plant threats and published that (Osawa and Inohara 2008). In terms of results, the volunteer activity expanded for the scientific activity.

After publishing the paper Osawa and Inohara 2008, the stakeholders of the program, including the participants, and park managers because interested in the scientific significance of these data. As a result, the framework of data management was established, and people who wanted to use the data could access with permission from park managers (Fig. 11.4b). I obtained permission and published one more paper (Osawa 2013).

In this case, transcend occurred by including the researcher (me) in the project (Fig. 11.4b). Specifically, it is successful to add the new goal without the occurrence of conflict. Finally, a new goal and collaboration type (passes) were added to the program, and this goal could be shared among the original participants (Fig. 11.4b). Importantly, the existing common goal was retained, which was also shared by participants (Fig. 11.4b). This study does not indicate that citizen science projects should include researchers. The first point is that citizen science projects could expand such as adding a new goal but maintaining a common, intermediate goal. The second point is that the motivation of participants could change through transcend.

## 11.6 Changes in Relationship and Collaboration Style in Citizen Science Projects

Even in a single case, the monitoring program in Hakone could provide several suggestions with regard to the characteristics of citizen science projects. The most important point is that the relationships among participants, including collaboration style, in the project do not remain constant but could change. In the Hakone case, many participants obtained a new interest after adding a new goal, that is, scientific publication, which led to new types of collaboration (Fig. 11.4). As a result, other participants became interested in academic publication (i.e., they were motivated to contribute to the literature). In fact, the activity of Hakone park volunteers, which included plant monitoring, contributed to other fields of academic research (Osawa et al. 2015, 2020) after this event. A few of the members endeavored to analyze their monitoring records. Thus, if the data policy of the project was strictly restricted, such that participants could decline the right to data use, then they may have felt disappointed at the time. Thus, the data policy in citizen science projects may require a certain degree of flexibility. This is one of the major differences of citizen science projects from ordinal research ones.

Sasaki et al. (2016) presented a diagram of the process of the changing interests of participants in citizen science projects (Fig. 11.5). According to the concept, several interests co-exist in citizen science projects, but individual interests could change with the support of a mediator (Fig. 11.5). Appropriate data treatment may play a role in each step of the interests of participants. In specific, there is motivation of participants for steps (b) and (c) in Fig. 11.5 for leisure and education; thus, they may not require the collected data they own and use. Alternatively, participants in step (d) in Fig. 11.5 may require the collected data they own and use. Citizen science projects frequently include different types of participants. This is one of the pre-conditions of a citizen science project. Importantly, however, the goals and

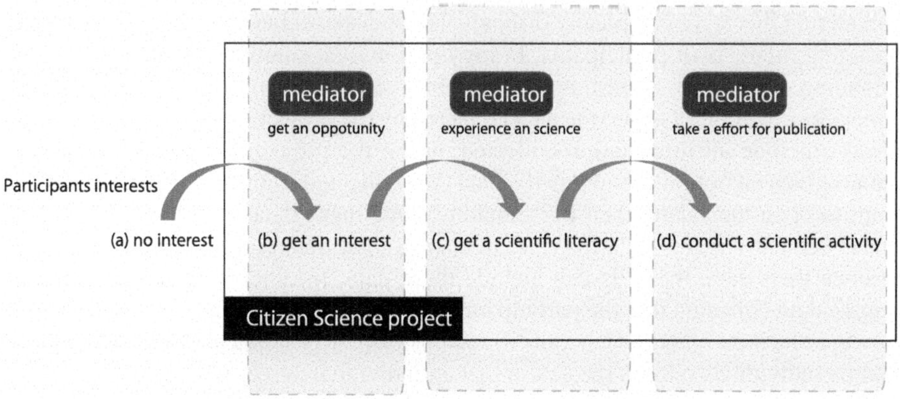


**Fig. 11.5** Process of changes in the interests of participants in citizen science projects. Sasaki et al. (2016) originally provided this diagram

motivations of participants in citizen science projects could change. Essentially, the data policy of citizen science projects, that is, the personal rights of the participants, should be flexible. In additions, PIs of citizen science projects should appropriately address such complex situations.

## 11.7 The Concept of Appropriate Data Policy for Citizen Science Projects

What then is considered an appropriate data policy in citizen science projects? Many studies indicated that a close relationship should exist between citizen and open sciences (Fritz et al. 2019; Heigl et al. 2019; Kondo et al. 2019; Haklay et al. 2021). Although arguments remain, the concept of open science includes open data (Kondo et al. 2019; Picaut et al. 2019; Dienlin et al. 2021), that is, data that can be shared and used with anyone and anywhere for any purpose (Osawa and Iwasaki 2016; Osawa 2019). Thus, data collected by citizen science projects become open data, which is one of its basic principles. In fact, many citizen science projects opened the collected data (Osawa and Wada 2016; Osawa et al. 2017; Suzuki-Ohno et al. 2021; Maruyama et al. 2022). This policy is appropriate from the perspective of open science, which essentially aims to increase the transparency of and trust in science (Allen and Mehler 2019; Heigl et al. 2019; Haklay et al. 2021). In addition, open data could promote data sharing and reuse, which can contribute to the development of science (Osawa and Iwasaki 2016; Osawa 2019). Essentially, scholars recommend the sharing of data collected by citizen science projects as open data. However, as discussed in this chapter, this issue within a project is slightly complex.

The basic method for establishing the data policy establishes a strict set of rules at the beginning of the project, and all participants observe this policy at all times. In general, research projects basically follow this style. However, citizen science projects are clearly different with an ordinally research project as described; thus, conducting this style is typically difficult. The indispensable point is addressing the changing interests of participants. In my opinion, one solution is categorizing and arranging data policy according to hierarchies such as for projects and for participants. Specifically, PIs open the data collected by the project; however, data may not always include all information collected during the project, because each participant is given an option of treating the data they collected. If other participants do not want to open their data, then PIs should exclude these data when the project data open. The participants hold the right to change their data policy; thus, if they want to open their data, then PIs can include these data. In other words, a project can release data collected that the participants agreed to open. This idea is similar to the concept of a *data bank*, which aims to establish a platform that supplies data deposited by individuals only to the extent that an individual agrees (Idota 2020).

Figure 11.6 demonstrates the hierarchy of the data policy framework. The project data policy should be stable basically as open data; thus, PIs should release the collected data periodically as open data. The released data should aggregate the data provided by the participants who agreed to policy A. The participants who refuse to

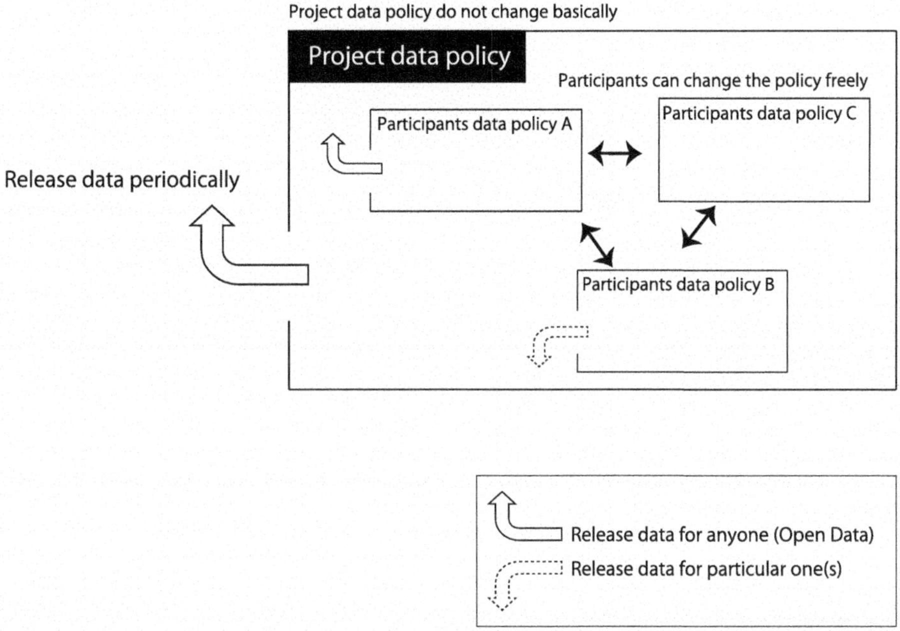


**Fig. 11.6** The basic idea of hierarchy in the data policy framework for citizen science projects. PIs can open project data with the agreement of participants, who also hold the right to change their data policy

open their data have two options, namely, policy B (sharing data for particular persons, typically project members) and policy C (they own their data and do not share even with project members). Of course, these policies could be more detailed such as reduce of precision (e.g., spatial grains of the species occurrence, changing a species identification accuracy from species to genus) in policy A or B, allow read-only, not use in policy B. Importantly, the participants can change their data policy anytime. However, retroaction is not allowed; in other words, once opened, the data should remain open, whereas closed data could be opened upon changing the data policy. This concept is one case that could address the changing interests of participants without infringing on their rights. The key issue of PIs is that citizen science projects should ensure an ethical equality for diversity, and the transmutative rights of participants in relation to data treatment should never be infringed upon.

## 11.8 Conclusion and Perspective

In this chapter, I addressed the ethical issues of citizen science projects with a particular focus on data policy and demonstrated one practical concept in this regard. In my opinion, citizen science projects are clearly different from ordinal research projects. Thus, a different data management policy is necessary. The essential difference is the variety in the attributes of participants, which could change within a project. Managing such complexity requires extreme efforts, which could not connect directly for academic publication. However, the PIs of citizen science projects should not disregard this complexity, because citizen science offers new pathways of cooperation among academics, citizens, politicians, and the private sector. Furthermore, it opens new paths for joining research (Vohland et al. 2019). With the inclusion of various actors, citizen science may need to establish new methods or alternative models for research activities (Vohland et al. 2019). Data policy is one of such important items that acts as a catalyst for the establishment of new strategies in science. Establishing adequate data policy procedures in a citizen science project may need new methods, such as co-creation with PI and participants, and gradual change with changing participant relationships. As such, practice and framing must be monitored, reflected on carefully, and improved.

**Acknowledgment** Author thanks to the Drs. Y. Suzuki-Ohno and Y. Ikkatai for discussion about the right of participants in a citizen science project. Also, author thanks to the member of Open Tram Science project especially Dr. Y. Kondo. This work was partially supported by the PARSEC project under Belmont forum.

## References

Allen C, Mehler DMA (2019) Open science challenges, benefits and tips in early career and beyond. PLoS Biol 17:1–14. https://doi.org/10.1371/journal.pbio.3000587

Buyx A, Del Savio L, Prainsack B, Volzke H (2017) Every participant is a PI. Citizen science and participatory governance in population studies. Int J Epidemiol 46:377–384. https://doi.org/10.1093/ije/dyw204

Dickinson J, Zuckerberg B, Bonter D (2010) Citizen science as an ecological research tool: challenges and benefits. Annu Rev Ecol Evol Syst 41:149–172. https://doi.org/10.1146/annurev-ecolsys-102209-144636

Dienlin T, Johannes N, Bowman ND et al (2021) An agenda for open science in communication. J Commun 71:1–26. https://doi.org/10.1093/joc/jqz052

Fritz S, See L, Carlson T et al (2019) Citizen science and the United Nations Sustainable Development Goals. Nat Sustain 2:922–930. https://doi.org/10.1038/s41893-019-0390-3

G7 (2019) Citizen science in the Internet era

Gajbe SB, Tiwari A, Singh RK et al (2021) Evaluation and analysis of data management plan tools: a parametric approach. Inf Process Manag 58:102480

Guerrini BCJ, Majumder MA, Lewellyn MJ, Mcguire AL (2018) Citizen science, public policy. Science 361:134–136

Haklay MM, Dörler D, Heigl F, et al (2021) What is citizen science? The challenges of definition. In: The science of citizen science, pp 13–33

Hayden CE (2012) Informed consent: a broken contract. Nature 486:312–314

Heigl F, Kieslinger B, Paul KT et al (2019) Toward an international definition of citizen science. Proc Natl Acad Sci U S A 116:8089–8092. https://doi.org/10.1073/pnas.1903393116

Idota H (2020) Ethical issues related to the distribution of personal data: case of an information bank in Japan. ETHICOMP, pp 326

Kobori H, Dickinson JL, Washitani I et al (2016) Citizen science: a new approach to advance ecology, education, and conservation. Ecol Res 31:1–19. https://doi.org/10.1007/s11284-015-1314-y

Kondo Y, Miyata A, Ikeuchi U et al (2019) Interlinking open science and community-based participatory research for socio-environmental issues. Curr Opin Environ Sustain 39:54–61. https://doi.org/10.1016/j.cosust.2019.07.001

Lee KA, Lee JR, Bell P (2020) A review of citizen science within the Earth sciences: potential benefits and obstacles. Proc Geol Assoc 131:605–617. https://doi.org/10.1016/j.pgeola.2020.07.010
Mahr D, Dickel S (2019) Citizen science beyond invited participation: nineteenth century amateur naturalists, epistemic autonomy, and big data approaches avant la lettre. Hist Philos Life Sci 41:1–19. https://doi.org/10.1007/s40656-019-0280-z
Marrone JA (2010) Team boundary spanning: a multilevel review of past research and proposals for the future. J Manag 36:911–940. https://doi.org/10.1177/0149206309353945
Maruyama N, Miyazaki K, Yasunaga E et al (2022) Long-term flowering records of herbaceous plants in Tokyo, Japan, during 1994—2015 through citizen science. Ecol Res 37:182–185
Matsubara A, Okoso M (2022) The technology of collaboration. Gakugei syuppan sya, Kyoto
Njue N, Stenfert Kroese J, Gräf J et al (2019) Citizen science in hydrological monitoring and ecosystem services management: state of the art and future prospects. Sci Total Environ 693:133531. https://doi.org/10.1016/j.scitotenv.2019.07.337
Osawa T (2013) Monitoring records of plant species in the Hakone region of Fuji-Hakone-Izu National Park, Japan, 2001–2010. Ecol Res 28:541–541. https://doi.org/10.1007/s11284-013-1049-6
Osawa T (2019) Perspectives on biodiversity informatics for ecology. Ecol Res 34:446–456. https://doi.org/10.1111/1440-1703.12023
Osawa T, Inohara S (2008) Understanding current status and factors of degradation of threatened plants in the Hakone region of the Fuji-Hakone-Izu National Park—Investigation using park volunteer research data. Jpn J Conserv Ecol 13:179–186
Osawa T, Iwasaki N (2016) Current status and future perspectives in open data in environmental sciences. Environ Inf Sci 44:35–40
Osawa T, Wada T (2016) Potential for citizen based observation for species distribution. Bird Res 12:R1–R8
Osawa T, Igehara G, Ito C et al (2015) A method for the early detection of vegetation degradation caused by Sika deer. Jpn J Conserv Ecol 20:167–179
Osawa T, Yamanaka T, Nakatani Y et al (2017) A crowdsourcing approach to collecting photo-based insect and plant observation records. Biodivers Data J 5:e21271. https://doi.org/10.3897/BDJ.5.e21271
Osawa T, Kato K, Tsujimoto A (2020) Lessons in eradication: successful reduction of the alien invasive plant Rudbeckia laciniata L. Jpn J Conserv Ecol 25:297–306. (in Japanese)
Picaut J, Fortin N, Bocher E et al (2019) An open-science crowdsourcing approach for producing community noise maps using smartphones. Build Environ 148:20–33. https://doi.org/10.1016/j.buildenv.2018.10.049
Rasmussen LM, Cooper C (2019) Citizen science ethics. Citiz Sci Theory Pract 4:1–3. https://doi.org/10.5334/cstp.235
Sasaki H, Ohnishi W, Osawa T (2016) Redefining "citizen science" from several perspectives. Jpn J Conserv Ecol 21:243–248
Silvertown J (2009) A new dawn for citizen science. Trends Ecol Evol 24:467–471. https://doi.org/10.1016/j.tree.2009.03.017
Suzuki-Ohno Y, Yokoyama J, Nakashizuka T, Kawata M (2021) Bee occurrence data collected in citizen science program "Hanamaru-Maruhana national census" in Japan. Ecol Res 36:1028–1036
Tanaka N (2005) Extraction of districts with intensive distribution of Reddata plants species in Kanagawa Prefecture. Bull Kanagawa Prefect Museum Nat Sci 34:47–54
Vaycna E, Tasioulas J (2013) Adapting standards: ethical oversight of participant-led health research. PLoS Med 10:1–5. https://doi.org/10.1371/journal.pmed.1001402
Vohland K, Weißpflug M, Pettibone L (2019) Citizen science and the neoliberal transformation of science—An ambivalent relationship. Citiz Sci Theory Pract 4
Welvaert M, Caley P (2016) Citizen surveillance for environmental monitoring: combining the efforts of citizen science and crowdsourcing in a quantitative data framework. Springerplus 5:1890. https://doi.org/10.1186/s40064-016-3583-5

# Chapter 12
# Citizen Science for Social Decision-Making: Avoiding Missteps and Unlocking Potentials

**Takeshi Osawa**

**Abstract** *Citizen science* that is involving the public in scientific inquiry has increased with an expected use in social decision-making as a democratized scientific method. Despite the success of citizen science projects, such as enhancing the public understanding of science and contributing to social well-being, the use of these projects in social decision-making is constrained due to an inherent issue, namely, *bias*. A critical challenge lies in the limited role of citizens in these projects, which impacts social decision-making. The representativeness of participants is a concern, because projects frequently attract a specific demographic, which can lead to results that are inapplicable to the broad population. Other issues include geographical and temporal representativeness, as well as the lack of accessibility and inclusion, which can exclude individuals without the necessary resources for participation. Additionally, a superficial involvement of citizens exists in the research process. Citizens are typically only involved in data collection and occasionally analysis, which leaves the crucial stages of design and interpretation to researchers. This scenario could also lead to results that are misaligned with community needs or values, which, hence, limits a genuine contribution to democratic and socially relevant decision-making. To realize the democratized scientific method, citizen science needs to extend beyond mere data collection and engage citizens in all aspects of the research process. By embracing these challenges, the future of citizen science could significantly impact society.

**Keywords** Community based · Decision-making · Demographic way · Inclusive · Social problem

T. Osawa (✉)
Graduate School of Urban Environmental Sciences, Tokyo Metropolitan University, Hachiouji, Tokyo, Japan



Y. Suzuki-Ohno (ed.), *Community Science in Ecology*, Ecological Research Monographs, https://doi.org/10.1007/978-981-97-0304-3_12

## 12.1 Introduction

In recent years, scientific evidence is one of the key factors of social decision-making (Lee 2010; Turner et al. 2022). Furthermore, the call for public involvement in social decision-making, such as establishing policy in line with democratic ideals, has increased (Rowe and Frewer 2000, 2005). *Citizen science* is typically heralded as a mechanism for greater public engagement with scientific research (Hollow et al. 2015; Hecker et al. 2019). This view leads to the assertion that citizen science may foster public input into social decision-making (Bonney et al. 2014; Hollow et al. 2015; Kobori 2022).

In fact, a number of previous studies suggested that citizen science projects can contribute to social decision-making in various ways. For example, Hollow et al. (2015) demonstrated that a citizen science project enabled the collection of a wide range of opinions, which helped in the discovery and definition of relevant issues on management and conservation policies for koalas (Hollow et al. 2015). Bonney et al. (2016) cited that citizen science can contribute to the public understanding of science and positively contribute to social well-being (Bonney et al. 2016). Tiago (2017) mentioned that guiding policy goals and decisions are one of the crucial factors of the appropriate design of citizen science projects. Therefore, citizen science projects hold the potential to engage the public in scientific research and contribute to social decision-making.

Scholars consider citizen science as a means of democratizing science and enhancing public engagement (Bonney et al. 2016; Hecker et al. 2019). Indeed, it offers significant potential in extending the reach and applicability of scientific research. To the best of our knowledge, however, discussions on the challenges related to citizen science for social decision-making remain few. Several studies have been conducted on the challenges for citizen science, especially on various types of *bias* (Dickinson et al. 2010; Sullivan et al. 2014; Kosmala et al. 2016). These issue also could limit its contribution to social decision-making. Therefore, in this column, I intend to construct the relationship between the challenges of citizen science and social decision-making. My contention is that although citizen science poses clear advantages for societal decision-making, its potential is more constrained than expected, which is mainly due to the essential challenges of citizen science. By leveraging these advantages and overcoming these challenges, citizen science holds the potential to significantly contribute to the process of societal decision-making.

## 12.2 Nature of Citizen Science and Conflicts in Democracy

Public engagement is one of the essential points of citizen science projects (Hollow et al. 2015; Hecker et al. 2019). These projects can range from simple observational studies to complex experimental designs; thus, they enable the public to contribute to various scientific disciplines (Sullivan et al. 2014; Bonney et al. 2014). When this principle is discussed along with democratic decision-making processes, one could infer that citizen science is regarded as a potent facilitator of endeavors in societal decision-making. However, the extent to which citizens can genuinely engage in the process significantly varies. This limited role presents several implications for social decision-making.

### *12.2.1 Bias and Representativeness*

The first challenge is the representativeness of participants in citizen science projects (Fig. 12.1a). In this aspect, demographic representativeness becomes an issue if citizen science projects attract participants mainly from specific age groups, socio-economic backgrounds, or levels of education (Hollow et al. 2015; Osawa 2018). This preference can lead to results that are inapplicable to the broad population. Geographical representativeness becomes a concern if projects attract participants from certain regions but not others, which leads to geographically skewed data (van Strien et al. 2013). For example, urban areas may be overrepresented, whereas rural or less accessible regions may be underrepresented (Dickinson et al. 2010). Temporal representativeness may occur if data collection is concentrated at specific times, such as weekends or holidays, because it may inaccurately represent a phenomenon over time (Callaghan et al. 2019). Accessibility and technology barriers could cause a lack of representativeness if a project requires specific skills or technology, which excludes those without access to these resources (Osawa 2018). Essentially, participants in citizen science project are volunteers; thus, maintaining representativeness for the target community may be difficult.

**Fig. 12.1** Inherent biases in citizen science

### 12.2.2 Accessibility and Inclusion

Certain issues emerge even among participants in citizen science projects. Not all individuals have equal access to participate in these projects (Fig. 12.1b). Individuals without access to necessary technology, education, or other resources may be excluded, which leads, once again, to potential bias and lack of representativeness (Chap. 11). The majority of citizen science projects involve the public in data collection and, occasionally, analysis. This contribution undoubtedly extends the reach of scientific research but frequently fails to engage citizens in other stages of research such as design and interpretation. Even in projects that claim to promote participatory science, researchers predominantly lead the formulation of research questions, methodology, and analysis (Dickinson et al. 2010). This scenario may reduce the ability of a project to reflect the interests, values, and local knowledge of the public (Fig. 12.1c). The final points of the results and their implications are typically dependent on researchers. This control may undermine the potential of citizen science to genuinely contribute to social decisions that align with the needs and perspectives of a community.

### 12.2.3 Scientific Accomplishments on Citizen Science Projects

Although citizen science projects could avoid or address bias using a circumspect design (Callaghan et al. 2019), preventing the essential bias frequently caused by researchers is difficulty. For this reason, scientific outputs are typically unable to reflect the collective opinion of these projects. When publishing accomplishments, one or a few professional researchers generally lead these projects (Fig. 12.1c). By maintaining control over study design and interpretation, researchers may inadvertently impose their perspectives and biases, which leads to results that are misaligned with the needs or values of the community. Participants in citizen science projects may hold various expectations or personal stakes in the research outcomes (Chap. 11). Thus, the superficial involvement of citizens without genuine empowerment may reduce citizen science to tokenistic participation, which undermines its potential to contribute to democratic decision-making. In reality, however, managing these expectations and ensuring that participants understand the objectives and limitations of these projects can be challenging. In fact, Ecklund et al. (2012) demonstrate that researchers hold the perception that the lack of scientific knowledge among citizens poses difficulty in communicating scientific discoveries (Ecklund et al. 2012). The advancement of mutual understanding between researchers and citizens (i.e., science communication is perceived as crucial for addressing this issue) and the discussion on this point are, indeed, progressing (Chap. 10; Kuehne and Olden 2015; Kondo et al. 2019).

## 12.3 Challenges of Citizen Science for Social Decision-Making

Although numerous citizen science studies have extensively noted these issues in bias, the primary focus has predominantly been on data bias (Dickinson et al. 2010; Kosmala et al. 2016) in which interpretative bias receives comparatively less discussion. Hence, the question emerges: how should the outputs of citizen science be leveraged in the context of societal decision-making?

Rigorous methodologies, extensive quality control, and in-depth analysis frequently characterize traditional scientific projects for social decision-making, which are typically conducted by professional scientists in the academic, governmental, and industry settings. This process produces high-quality, peer-reviewed knowledge that serves as a solid foundation for informed social decision-making. However, such projects may not always reflect the diversity of community perspectives, may be limited by their specific focus or availability of resources, and may oftentimes be disconnected from on-the-ground reality. Alternatively, citizen science projects are considered to engage the public in scientific investigation, which offers an opportunity for increased community participation and the collection of data at scales or resolutions that may be unfeasible for professional scientists (Bonney et al. 2014; Callaghan et al. 2019). These projects can capture a wide range of perspectives and foster public understanding and engagement in science. However, as previously discussed, the advantages attributed to citizen science are not always guaranteed. Importantly, when scientists are largely involved in the aggregation of collected data, the scientific achievements that can be garnered may not necessarily be inferior to those of traditional science. Although both traditional and citizen science projects approach discovery and understanding in distinct ways, they are united in their shared pursuit of scientific knowledge and accomplishments. Additionally, they share weak points such as difficulty in reflecting diverse perspectives and limitation in resources. In summary, no essential differences exist between the two forms of science, at least in terms of scientific accomplishment, which could contribute to social decision-making (Fig. 12.2).

Conversely, citizen science poses certain specific advantages. As widely discussed, this aspect primarily involves the process in which stakeholders that is, the citizens, are directly involved in problem-solving and issue identification. Despite various expectations, the role of science is to provide objective responses. However, the direct involvement of citizens in this process is a challenge for traditional science. In this respect, the application of the outcomes of citizen science to societal decision-making could potentially bear significance, because it realizes an endeavor that was previously unattainable (Fig. 12.2). Thus, in social decision-making, recognizing the differences between traditional and citizen science, instead of avoiding their dichotomization, would be beneficial.

Social decision-making should be based on a comprehensive understanding that includes the rigorous insights offered by the scientific method and the inclusive perspectives provided by citizens. Notably, these two approaches can be

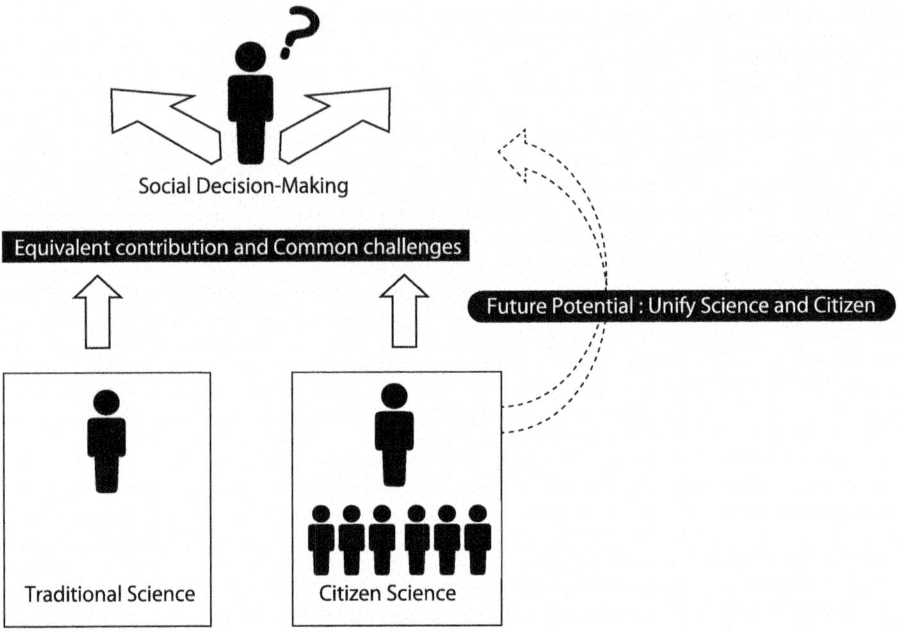


**Fig. 12.2** Contributions of traditional and citizen science to social decision-making and the potential of citizen science

complementary, and integrating them can lead to more robust and inclusive decision-making processes. From this perspective, citizen science possesses the potential to encompass the critical components of societal decision-making (Fig. 12.2). However, realizing this potential will likely be a significant challenge for citizen science in the future.

## 12.4 Conclusion and Perspective

In conclusion, this column intends to make two assertions. First, no essential difference exists between traditional and citizen science in terms of outcomes; accordingly, no disparity exists in their contribution to societal decision-making. Second, citizen science ensures an advantage in that it permits the participation of citizens in the knowledge acquisition process and possesses the potential to encompass the critical components of societal decision-making. To truly unlock the potential of citizen science in the formation of social decisions, a radical rethinking of participation is required. This process entails going beyond mere data collection to genuinely engaging citizens in all aspects of scientific inquiry, that is, from the formulation of a problem to the application of the result as well as its interpretation and publication. Without this shift, citizen science will likely remain an extension of professional scientific practice instead of a transformational force for a more democratic, socially relevant decision-making.

Although citizen science offers a tantalizing prospect for bridging the gap between scientific research and social decision-making, the existing practices and paradigms frequently constrain its true potential. To transform citizen science into a genuine force for democratic, socially relevant decision-making, a concerted effort must be exerted to evolve beyond the current limitations. This aspect requires a reimagining of participation, a commitment to genuine collaboration, and vigilant monitoring to ensure that citizen science is not merely an extension of professional practice but a transformative approach that genuinely includes and values the voices and contributions of citizens. The future of citizen science lies in embracing these challenges and, in so doing, realizing its potential to impact not only science but also society as a whole.

**Acknowledgments** Author thanks to the Dr. Y. Suzuki-Ohno and Dr. Y. Ikkatai for discussion about the issues on the relationship between social issue and citizen science project. Author used ChatGPT 4.0 (https://chat.openai.com/) for English editing.

## References

Bonney R, Shirk JL, Phillips TB et al (2014) Next steps for citizen science. Science 343:1436–1437. https://doi.org/10.1126/science.1251554

Bonney R, Phillips TB, Ballard HL, Enck JW (2016) Can citizen science enhance public understanding of science? Public Underst Sci 25:2–16. https://doi.org/10.1177/0963662515607406

Callaghan CT, Rowley JJL, Cornwell WK et al (2019) Improving big citizen science data: moving beyond haphazard sampling. PLoS Biol 17:e3000357. https://doi.org/10.1371/journal.pbio.3000357

Dickinson JL, Zuckerberg B, Bonter DN (2010) Citizen science as an ecological research tool: challenges and benefits. Annu Rev Ecol Evol Syst 41:149–172

Ecklund EH, James SA, Lincoln AE (2012) How academic biologists and physicists view science outreach. PLoS One 7:e36240. https://doi.org/10.1371/journal.pone.0036240

Hecker S, Wicke N, Haklay M, Bonn A (2019) How does policy conceptualise citizen science? A qualitative content analysis of international policy documents. Citiz Sci 4:32. https://doi.org/10.5334/cstp.230

Hollow B, Roetman PEJ, Walter M, Daniels CB (2015) Citizen science for policy development: the case of koala management in South Australia. Environ Sci Pol 47:126–136. https://doi.org/10.1016/j.envsci.2014.10.007

Kobori H (2022) Advocating citizen science: changing science, education, and society through 'Personal' and 'Collective' endeavors. Bun-ichi sogo syuppan (in Japanese)

Kondo Y, Miyata A, Ikeuchi U et al (2019) Interlinking open science and community-based participatory research for socio-environmental issues. Curr Opin Environ Sustain 39:54–61. https://doi.org/10.1016/j.cosust.2019.07.001

Kosmala M, Wiggins A, Swanson A, Simmons B (2016) Assessing data quality in citizen science. Front Ecol Environ 14:551–560

Kuehne LM, Olden JD (2015) Lay summaries needed to enhance science communication. Proc Natl Acad Sci U S A 112:3585–3586. https://doi.org/10.1073/pnas.1500882112

Lee YC (2010) Developing decision-making skills for socio-scientific issues. J Biol Educ

Osawa T (2018) ICT-driven possibilities in citizen-participation ecological surveys, and the urgent challenges toward establishing a sustainable system. In: Ecology innovated by information and communication technology: accelerating open data and open science. The Society for the Study of Species Biology (in Japanese), pp 1–15. https://www.speciesbiology.org/publications/e-book/2.html

Rowe G, Frewer LJ (2000) Public participation methods: a framework for evaluation. Sci Technol Hum Values 25:3–29. https://doi.org/10.1177/016224390002500101

Rowe G, Frewer LJ (2005) A typology of public engagement mechanisms. Sci Technol Hum Values 30:251–290. https://doi.org/10.1177/0162243904271724

Sullivan BL, Aycrigg JL, Barry JH et al (2014) The eBird enterprise: an integrated approach to development and application of citizen science. Biol Conserv 169:31–40

Tiago P (2017) Social context of citizen science projects. In: Analyzing the role of citizen science in modern research. IGI Global, pp 168–191

Turner S, D'Lima D, Sheringham J et al (2022) Evidence use as sociomaterial practice? A qualitative study of decision-making on introducing service innovations in health care. Public Manag Rev 24:1075–1099. https://doi.org/10.1080/14719037.2021.1883098

van Strien AJ, van Swaay CAM, Termaat T (2013) Opportunistic citizen science data of animal species produce reliable estimates of distribution trends if analysed with occupancy models. J Appl Ecol 50:1450–1458. https://doi.org/10.1111/1365-2664.12158

# Part IV
# Changing Environments with Participants

# Chapter 13
# The Importance of Dialogue Between Science and the Public Shown in the Invasive Bumblebee Problem

**Koichi Goka**

**Abstract** For ecologists, it is "common sense" to thoroughly regulate and control invasive species from the perspective of biodiversity conservation. However, this common sense does not necessarily apply to everyone in society. People in society have various positions and jobs, and there is no universal common sense or justice that applies to all. We scientists assess the risks based on scientific data and advocate for the necessity and methodology of risk management for invasive species. However, if these scientific processes are not accepted by society, the effectiveness of invasive species risk management will be lost.

The European bumblebee, *Bombus terrestris*, has been imported as an agricultural pollinator and has greatly contributed to tomato production in Japan. However, at the same time, it has caused ecological impacts as an invasive species. We have scientifically clarified the actual ecological damage caused by this species and, in collaboration with stakeholders such as the government, citizens, and makers, devised methods to continue agricultural utilization while managing the risks caused by this species.

Furthermore, when we were developing chemical control methods using insecticides for the naturalized populations of the European bumblebee, we were able to encourage efficacy testing by obtaining the cooperation of the public. However, during the implementation phase at the control sites, due to inadequate risk communication, we faced strong opposition from some local residents regarding the use of insecticides for chemical control. From this failure, we learned important points about risk communication in chemical control.

**Keywords** Invasive alien species · *Bombus terrestris* · Ecological risk · Chemical control · Insecticide · IGR · Risk communication

---

K. Goka (✉)
National Institute for Environmental Studies, Tsukuba, Ibaraki, Japan
e-mail: goka@nies.go.jp


Y. Suzuki-Ohno (ed.), *Community Science in Ecology*, Ecological Research Monographs, https://doi.org/10.1007/978-981-97-0304-3_13

## 13.1 Introduction

The topic here is about the dialogue between science and the public about a bee called the bumblebee. Bumblebees are cute little bees with hairy bodies. In Japan, it is difficult to see bumblebees in urban areas, but in the suburbs, you can often see the queen bee, after waking from hibernation, flying around collecting pollen and nectar around the time azaleas are in bloom. In Japan, there are 20 native species of bumblebees, and it is considered that their pollination plays a pivotal role in supporting the rich flora of this country. On the other hand, a large number of exotic species of bumblebee have been imported into Japan since the 1990s. The fact that this alien bumblebee has become established and is expanding its distribution in Japan has become a major conservation biology problem. Here, we would like to review the process of collaboration and confusion between scientists and citizens over the management of invasive bumblebee and discuss the importance of dialogue and consensus between science and citizens in the control of invasive species (Fig. 13.1).

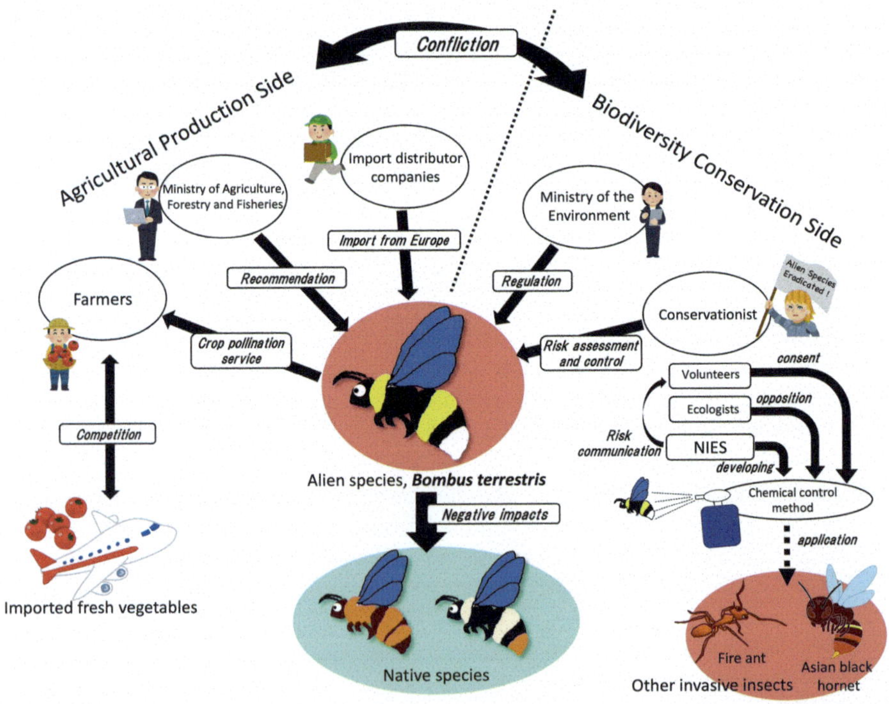


**Fig. 13.1** Collaboration and confusion between scientists and citizens over the management of invasive bumblebee

## 13.2 The European Bumblebee Introduced to Japan

The European bumblebee *Bombus terrestris* is a bee species widely distributed in Europe (Fig. 13.2). This alien bumblebees have been imported to Japan to serve as pollinators for tomatoes grown in greenhouses and glasshouses. Pollinators carry pollen from flower to flower to assist in pollination, and flowers pollinated by bees are able to produce fruit.

The most famous pollinator for agricultural crops is the European honeybee *Apis mellifera*. The species has been imported and used in Japan since the Meiji era (1868–1912). However, tomato flowers do not produce nectar, and pollen cannot be collected without vibrating the anthers (the pollen-producing organ at the tip of the stamen), so *Apis mellifera* cannot be used. Therefore, until bumblebees were introduced, farmers either carried pollen by hand or forced flowers to produce fruit without pollination by pouring a chemical called "plant growth hormone" on the flowers. This was quite hard work and a heavy burden for the farmers.

Then, the use of bumblebees was devised. Bumblebees visit flowers that do not produce nectar to collect pollen. Furthermore, bumblebees are very good pollinators for tomato flowers because they collect pollen by vibrating their large bodies, which aids pollen dispersal and increases pollination efficiency. Bumblebees were first commercialized in Europe, where greenhouse cultivation is popular. In the 1980s, artificial breeding methods for bumblebees were established, factories were

**Fig. 13.2** A worker of the European bumblebee *Bombus terrestris* visiting the flower of white clover *Trifolium repens* in the Hokkaido field (Photo by Daito M, National Institute for Environmental Studies)

built for mass propagation, and "artificial nests (colonies) of the European bumblebee" began to be actively marketed in the 1990s.

An artificial colony of the European bumblebee is a portable plastic case in which the bumblebees are allowed to build a hive. The case has an open-and-shut hole for worker bees to enter and exit. When the hive is placed in a plastic greenhouse or glasshouse where tomatoes are grown and the hole is opened, the worker bees emerge from the hive, horber around the greenhouse, collect pollen, and return to the hive. This pollen collection promotes pollination.

The commercialized European bumblebee rapidly came into use throughout Europe, and Japan began introducing them in 1992. At the time of its introduction, the commodity colony had an annual distribution of only 4000 colonies. Since then, the product has continued to sell well, and since the 2000s, annual distribution has grown to 70,000 colonies. Today, the European bumblebees are responsible for a large portion of Japan's tomato production.

## 13.3 Bumblebee Special Tomato

The introduction of the European bumblebee has greatly contributed to tomato production in Japan in three ways. The first is the improvement of productivity. As mentioned earlier, before the introduction of bumblebees, farmers had to pollinate tomatoes by hand or treat each flower with the chemical hormone-like compound, 2–4 D. Thanks to the bumblebees' help in pollination, farmers' workload has been greatly reduced. The easier workload has allowed farmers to increase the number of greenhouses and the scale of production.

Second, it improves quality. Bumblebee-pollinated "bumblebee tomatoes" are juicy and sweet, with well-developed seeds and packed fruits. Third, it improves safety. Because live bumblebees are placed inside vinyl and glass houses, the amount of pesticides applied in them must necessarily be reduced. Pesticide-free and pesticide-saving cultivation using biological pesticides will be promoted, and as result, safe and secure tomatoes will be available to consumers. In other words, the use of the European bumblebee has brought great benefits to both farmers and consumers.

## 13.4 Is the European Bumblebee an Invasive Alien Species?

While the European bumblebee is indeed a valuable biological asset, concerns have arisen among biologists ever since its introduction to Japan was contemplated. Many biologists claimed that the European bumblebee is an "invasive alien species" that could potentially jeopardize the native ecosystem. The basis for this concern lies in the fact that the European bumblebee evolved on the continent and possesses strong competitive traits. If it were to establish itself in Japan's natural habitats, there would be a significant risk of it displacing the native Japanese bumblebee and dominating the pollination ecosystem.

In fact, there have been many reports in the world that in areas where the European bumblebee has been introduced, and has become feral, the alien bumblebee has reduced the number of native bumblebees. There are more than 20 species of native bumblebees in Japan. These native bumblebee species are thought to co-evolve with Japanese plant species and form unique symbiotic relationships with those. The invasion of the European bumblebee has the potential to stir up these bee-plant relationships and even negatively impact the plant ecosystem.

In 1998, the European bumblebees that had escaped from plastic greenhouses in Hokkaido were found to be nesting in the field. The number of feral nests increased over the years. It also became clear that the number of native bumblebees that originally inhabited the area where the number of the European bumblebees had increased was decreasing. The European bumblebee has certainly begun to make negative impacts into the Japanese ecosystem.

## 13.5 The Use of the Commercial Bumblebee Should Be Banned

Under these circumstances, the Invasive Alien Species Act was enacted in 2004 by the Ministry of the Environment. Many ecologists argued that the European bumblebee should be regulated under the Invasive Alien Species Act, and both importation and use of the alien bee should be banned. There was even a debate in the Diet regarding the regulation of this bee. The problem of the European bumblebee, however, is not so simple that it can be simply solved by the legal regulation. The alien bee is a "beneficial insect" imported for agricultural purposes, and it was the Ministry of Agriculture, Forestry, and Fisheries that decided to import the species. Should an alien species that contributes to agricultural production be easily made subject to regulations on the grounds that it has a negative impact on the environment? The alien bee was caught between two ultimately conflicting propositions: agricultural production and environmental conservation.

Japan is a major food importing country, with total food self-sufficiency rate on a calorie basis (domestic heat supply per person per day/domestic heat supply per person per day) of less than 40% (from 1970s to 2023). This figure is very low compared to other developed countries. We Japanese must take the significance of this figure very seriously.

As the Japanese people repeatedly consumed and imported large quantities of food, rice and fresh vegetables were the last bastions of food self-sufficiency in Japan. However, this situation has changed since the late 1990s. The self-sufficiency rate of raw vegetables began to decline and is now said to be less than 90%. Instead, imports of vegetables are increasing. The largest import partner is our neighbor China.

The increase in imports of fresh vegetables is not only simply due to a drop in domestic production in Japan but also to the liberalization of international trade. Since the establishment of the World Trade Organization (WTO), liberalization of trade in agricultural products has also been pushed forward on a global scale. Furthermore, the development of large cargo aircraft and means of transportation such as high-speed transport ships has further facilitated international trade in fresh vegetables.

Countries that had protected their own vegetable production by imposing tariffs on imported vegetables, including Japan, are now obliged to eliminate tariffs and actively import vegetables from abroad as long as they are members of this organization. In particular, Japan is a major exporter of industrial products, including automobiles, and is in a position to promote the import of agricultural products in order to have foreign countries buy these machinery products and parts. If this trend continues, Japanese agriculture may decline rapidly.

In order for Japanese agricultural products to compete with imported agricultural products in the free market economy, labor saving is the first thing that must be considered. We must consider how to produce vegetables efficiently at low cost and how to win the price competition. The next key for international competition in agricultural products is to achieve high quality and high value added. We need to produce and offer fresh and tasty vegetables, aiming to differentiate them from foreign-produced vegetables. Another strategy is to sell safety and security. We must try to compete with imported vegetables by eliminating or reducing the use of pesticides.

Tomato imports have also increased dramatically since the late 1990s. Most of them are imported from South Korea. If nothing is done, Japan's tomato production will lose out to imported vegetables. Under such circumstances, in response to the call from the production sites for "help from insects," the European bumblebee appeared on the scene as a knight on a white horse. These bees brought labor savings, higher quality, and safety and healthy to tomato production. Can we really regulate insects that are saving Japan's agricultural production in a critical situation in the name of ecosystem protection? The issue of the European bumblebee was an important one for the future of Japanese agriculture.

## 13.6 Pathways to a Solution

When the Invasive Alien Species Act was soon to be enacted and discussions about the fate of the European bumblebee began to flourish, I was opposed to the enactment of this law. As an ecologist, this may sound preposterous, but to me, there were still many obstacles to managing invasive alien species within the framework of a law at this point in time. The most important reason was that the ecologically important concepts of "biodiversity" and "invasive alien species" were far from being well understood by the general public.

In anticipation of COP10 in Nagoya in 2010, the government conducted a survey of the general public in 2008, and to some shocking extent, more than 60% of the respondents said they had never even heard of the term "biodiversity." In other words, the ecological common sense that "an increase in alien species will have a negative impact on Japan's ecosystem" is not commonly accepted in society. Although it is necessary to establish common knowledge starting from the definition of alien species, education and dissemination still require more time. In this situation, I thought that even if laws were suddenly enacted to regulate invasive alien species, people would not understand what it means. However, in reality, the law was already in the process of being drafted, and it was us, a group of researchers called ecologists, who have sincerely hoped for the enactment of this law and worked hard to persuade the government.

Since the law had been decided to be enacted, we also needed to find a solution to the problem of the European bumblebee. As the possibility of the European bumblebee being regulated by law increased, concerns among farmers grew, and the Ministry of Agriculture, Forestry and Fisheries could not stay silent. In some cases, agricultural production could take priority, and regulation of the European bumblebee might be abolished. However, if that were the case, the European bumblebee would continue to be vilified, and the farmers who use it might suffer from reputational damage. On the other hand, completely banning the use of the European bumblebee would inevitably have a negative impact on tomato production.

## 13.7 Is There Any Way to Find a Productive Solution?

I, who had been at odds with the Ministry of the Environment over the pros and cons of enacting the law, began negotiations with them. As I read through the draft of the law, I came up with an idea. The law prohibits the import and breeding of designated invasive alien species, but as an exception, it states that "import and breeding for research purposes can be done with the permission of the Minister of the Environment." I wondered if "agricultural production purposes" might also be included in the "research purposes."

Furthermore, the law requires that closed breeding facilities be used to obtain permission for the breeding of designated invasive species, but I thought that vinyl houses that are closed with nets or other means could also be recognized as breeding facilities. After proposing this to the Ministry of the Environment, they approved it.

As a result, we decided to work with the Ministry of Agriculture, Forestry and Fisheries to create a special exemption that would allow the use of the European bumblebee as long as it is used in vinyl houses where precautions have been taken to prevent the bees from escaping, with permission from the Minister of the Environment, despite the regulation of this bee as a designated invasive species.

## 13.8 Launch of Project B

We have set a goal toward the regulation of the European bumblebee and formed a project team consisting of ecological researchers from universities, companies selling the commercial bumblebees, as well as oversight agencies such as the Ministry of the Environment and the Ministry of Agriculture, Forestry and Fisheries. The first meeting of the project was held in Biratori-cho, Hokkaido, which is a major production area of the bumblebee-made tomato. At this time, we decided to call the project "Project B," taking the "B" from both the Bumblebee and the town of Biratori.

In this project, when we were preparing to bring back the European bumblebee under the Alien Species Act, we were told by officials from the Ministry of Agriculture, Forestry, and Fisheries that we needed to provide "enough evidence to convince the Ministry" that the naturalized European bumblebee was truly having a negative impact on the ecosystem.

Therefore, we decided to thoroughly scientifically verify the negative impact caused by the European bumblebee through a research project led by the National Institute for Environmental Studies.

As a result, by 2006, it had become clear that the naturalized European bumblebee was having the following impacts on native species:

1. Many species of bumblebees make nests in underground cavities, such as abandoned rodent nests. The European bumblebee is competing for these nesting sites with native bumblebees and squeezing them out of their habitat.
2. Male European bumblebee can mate with queens of native species, causing reproductive interference and hindering the reproduction of the native species.
3. The European bumblebee has introduced parasitic mites from Europe, which are spreading to native bumblebees.
4. They engage in nectar-robbing behavior toward the Hokkaido native plant species *Corydalis fumariifolia* subsp. *azurea*, Ezoengosaku, which involves drilling holes in the flowers to steal nectar without pollinating, negatively affecting seed production.

Among these impacts, the problem of (2) reproductive interference has deeply resonated with many policymakers and has had a significant effect on raising awareness for them that the European bumblebee is a serious invasive alien species that threatens the survival of native bumblebees.

## 13.9 Cooperation of Farmers

The next challenge was to consider whether there were any means to avoid the negative impacts of the European bumblebee and to establish a management method. In order to do so, it was necessary to first raise awareness among farmers, who are the users, about the fact that the European bumblebee is an invasive species and why invasive species should not be released into the wild and to promote the correct usage methods.

To that end, we started by holding "study sessions" in Biratori-cho in Hokkaido and then expanded this initiative throughout the country to gather farmers, agricultural cooperatives, and other agricultural stakeholders to explain and increase understanding about why the European bumblebee is being regulated under the Invasive Alien Species Act.

In order to verify the effectiveness of preventing the escape of the European bumblebee by covering the entire vinyl greenhouse with a net, tomato farmers in Biratori-cho conducted a large-scale test by covering all the vinyl greenhouses in the town with their own money. Although it was a significant burden for the farmers, they aimed for the coexistence of tomato production and environmental conservation by understanding the necessity and importance of managing invasive species through study meetings.

This netting had a dramatic effect. Before the netting, in Biratori-cho where a large amount of European bumblebees were used, many of them escaped from the greenhouse every year and were found all over the town. However, after the netting, the number of European bumblebees caught in the surrounding outdoors significantly decreased. Therefore, netting the greenhouse was considered an extremely effective method for preventing the bees from escaping.

The Ministry of the Environment and the Ministry of Agriculture, Forestry and Fisheries decided to promote the unified method of covering with a net nationwide. In addition, it was decided to properly dispose of used beehives instead of leaving them outside before disposal.

With the accumulation of evidence on the negative impact of the European bumblebee on the ecosystem and the establishment of means to avoid the negative effects, the Ministry of the Environment designated the European bumblebee as a specified invasive species in 2006, which requires permission from the Minister of the Environment for agricultural use. Under the government's management, a groundbreaking example of using invasive species that are harmful to the environment was created here.

It can be said that the resolution of the dilemma of regulating beneficial insects and achieving a soft landing was the result of efforts by all stakeholders, including ecologists, policymakers, sellers, and users, namely, farmers, who engaged in dialogue.

## 13.10 Remaining Challenge

The European bumblebee has been designated legally as an Invasive Alien Species, and if farmers purchase its commercial colony and use it for tomato cultivation, they must obtain permission from the Ministry of the Environment. With this regulatory system, the situation where European bumblebees escape from the greenhouse into the outdoors has been significantly improved. However, on the other hand, control of populations that have already escaped and become established in the wild has been slow.

One reason for the slow progress of control is the lack of clarity and rationality regarding the responsible entity for control. Since the European bumblebee is regulated by the Invasive Alien Species Act, the Ministry of the Environment (national government), which is responsible for the law, must take responsibility for controlling the wild population. The government will request cooperation from local governments for control. However, local governments have limited budgets, making it difficult to implement sufficient control measures.

First of all, this alien bumblebee is an industrial product, and if some damages or destructions occurs as a result of the product, the manufacturing company should compensate for the damages under the manufacturer's responsibility. In this case, compensation should mean nothing other than promptly eradicating the naturalized European bumblebee and restoring the ecosystem to its natural state. However, unfortunately, such discussions have not been made, and ultimately, in Hokkaido where the European bumblebee has become wild, the Hokkaido government has been at the forefront of control, recruiting volunteers and working on European bumblebee control with a small budget in accordance with the criteria of the Invasive Alien Species Act.

The other reason why the alien bumblebee control did not progress was the lack of accurate control technology development. The control method implemented by the volunteers recruited by the Hokkaido government was a physical control method using insect nets to capture wild individuals. Volunteers were gathered by region starting in 2007, and the overwintering queen and summer worker bees were captured by them. As a result, 20,000 to 50,000 individuals of the European bumblebee were captured and killed in a year. However, despite these capture efforts, it was not enough to apply sufficient capture pressure to the reproductive potential of the European bumblebee, and according to the capture data until 2012, the capture numbers did not decrease significantly.

Therefore, our research team at the National Institute for Environmental Studies embarked on the development of scientific technology to reliably reduce the population density of the European bumblebee.

## 13.11 Trial of Chemical Control of the European Bumblebees

Before I began working at the National Institute for Environmental Studies, I belonged to the agricultural chemical research division of a chemical manufacturer, where I worked on developing insecticides. Using the knowledge and experience I gained during that time in pesticide science, I came up with ideas for chemically controlling invasive bumblebees.

In social insects like bumblebees, no matter how many worker bees are captured, as long as the nest is alive, the workers will be quickly replenished and the nest will be maintained. While capturing a hibernating queen bee before she begins to build the nest may seem effective in stopping nest production, in the case of the European bumblebee, hundreds of new queens are produced from a single nest. In other words, new queens are overproduced, and they compete with each other for nest sites in the spring, evolving to eliminate weaker genes. Therefore, in situations where wild nests are already widespread, capturing new queens with only moderate pressure will only have the effect of thinning out the queen population.

The essential point of control is to prevent the production of new queens from existing wild nests by eliminating the queen inside the nest as well as the larvae that will develop into new queens. Therefore, the method I came up with is to spread insecticides on the worker bees that are flying outside, attaching the chemicals to their bodies or the pollen they carry and then letting them carry the chemicals into the nest, exposing the queen and larvae inside the nest.

However, with regular insecticides, there is a risk that the sprayed worker bees will become immobilized before they reach the nest. So, I came up with the idea of using insect growth regulator (IGR) agents. IGR agents inhibit the enzyme that synthesizes chitin, which forms the exoskeleton of insects, thus preventing the insects from molting and causing them to die. The IGR agent inhibits the molting of larvae and also shows an effect of preventing the hatching of eggs inside the queen. On the other hand, it is harmless to adult workers that do not molt (Fig. 13.3).

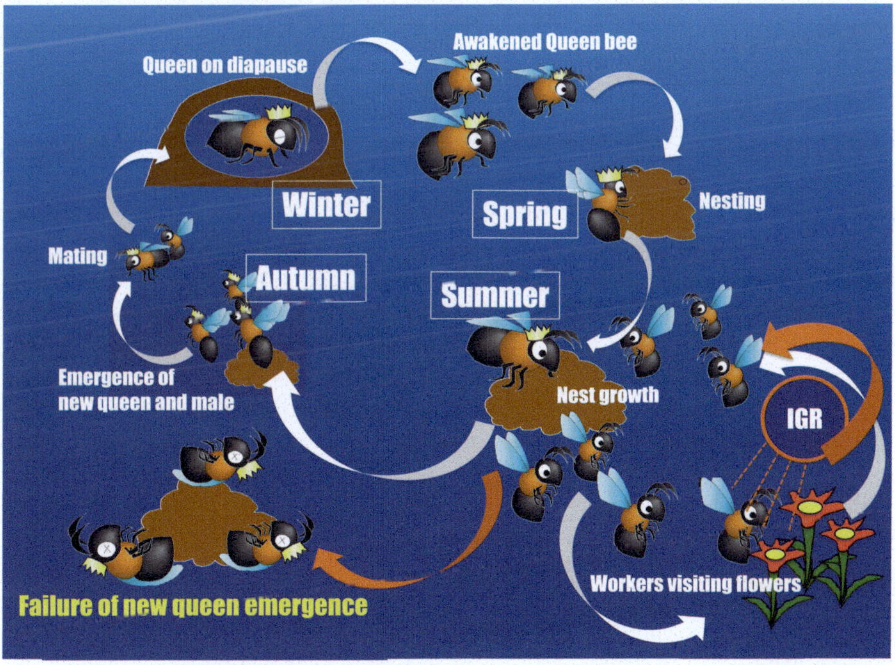


**Fig. 13.3** Principles of chemical pest control utilizing the life history of the bumblebee

We conducted indoor and greenhouse-level testing using the IGR and confirmed its effectiveness against colonies of the European bumblebee; in colonies where bees treated with the chemical returned, the production of new queens was suppressed. The next stage of chemical control development was efficacy testing against wild nests.

When we presented this chemical control method at the Entomological Society Conference held in Hokkaido, we were interviewed by the local newspaper. As a result, we received offers of cooperation from residents of Hokkaido for outdoor testing of chemical control. In the city where the residents lives, there are cases where the European bumblebees nest under the floors of houses, and they suggested that we could conduct chemical control tests using those nests.

We were able to conduct a pesticide test with the cooperation of local residents using multiple colonies of the European bumblebee discovered in residential areas. We applied the pesticide to some of the worker bees that emerged from the nest and asked the residents to observe the number of worker bees and new queens that emerged from the nest thereafter. As a result, it was confirmed that in the colonies not exposed to the pesticide, new queens were produced at the end of summer, whereas in the treated colonies, the number of new queens flying was zero, demonstrating the effectiveness of the pesticide.

Bumblebee workers, during the spring to summer period, collect nectar and pollen from flowers and bring them back to the nest to nourish the larvae. By applying IGR agents to the worker bees, it is possible to introduce these agents into the nest, exposing the larvae to them, ultimately halting their growth and preventing the production of the next generation.

## 13.12 Barriers to Consensus Building

We then proceeded to prepare for the next stage of the control method development, which was pesticide testing on the European bumblebee populations in the field. We select a certain peninsula that we had been tracking the distribution expansion of the European bumblebee as a candidate test site. In this area surrounded by the sea, we planned to prevent the naturalized nests from spreading by capturing and spraying the insecticide on the foraging worker bees of the European bumblebee and then releasing them. We then evaluated the effectiveness of the insecticide treatment based on the flight numbers of the worker bees in the test area in the following year.

Prior to the test, we prepared a meeting to explain the test plan to the town hall and residents who manage the peninsula and to obtain their agreement. However, just before the meeting, I, who was the head of the testing and development team, fell seriously ill and had to be hospitalized for 6 months. As a result, I entrusted another researcher of bumblebees to explain the plan at the meeting. However, the meeting went down in flames, and it caused a huge backlash on social media. The local residents strongly opposed the use of insecticide, leading to the failure of the agreement. The chemical control project hit a snag.

I received the report about the meeting while lying in the hospital bed and tried to analyze the causes of the failure. The fundamental reason was considered to be a lack of mutual understanding between the residents and the scientist. In particular, the scientist failed to address the residents' "concerns" about insecticide use and perceived their concerns as "ignorance" or "misunderstanding," which was the problem. "Scientific common sense," which draws conclusions based on numerical data, is not always common sense for the general public.

After being discharged from the hospital, I started a public awareness campaign about chemical control tests. I made sure to keep the following points in mind during the campaign:

1. Risk assessment of pesticides based on scientific data as an expert in pesticide science and disclosure of the data
2. Humble attitude and careful explanation
3. Transparency and cooperation from method planning to development and implementation (incorporating opinions and proposals from residents)
4. Explanation of adaptive management, which involves stopping the test and investigating the cause to provide feedback if any problems are observed after the test has started
5. Character-building for credibility, trust, and comprehensibility

As a result, understanding of chemical control has progressed among citizens, and local governments have become more positive about establishing trial projects. However, local governments still find it difficult to secure the budget, and so far, they have not been able to carry out large-scale trials. In addition, there is still strong opposition from experts in the field of ecology against the use of insecticides, and the chemical control technology for the European bumblebee is at a standstill, just before practical use.

I do not believe that the risk of chemical control is zero. However, if we can enlist the help of the volunteer group that has been cooperating in the capture of the European bumblebees for many years, we can selectively capture the European bumblebees with a high degree of probability from the bumblebee communities in the field, and we can apply insecticide with minimal ecological impact. I still question whether it is an appropriate attitude for researchers to simply assert the danger of pesticides without proposing alternative effective control methods or without sufficient knowledge of pesticide science.

Currently, our research team is focusing on emergency control against the following more dangerous invasive insects, which has delayed the promotion of European bumblebee control. However, we aim to continue to uphold the roadmap for the chemical control that has been created through collaboration among citizens, administration, and researchers, and ultimately achieve its realization.

## 13.13 Application and Success of Chemical Control

During the development of chemical control for the European bumblebee, a new invasive alien insect species appeared in Japan, and our research team had to respond urgently to their control.

In 2010, the invasion and establishment of the Argentine ant, *Linepithema humile*, a species native to South America, were confirmed in Tokyo, and in 2012, the invasion and establishment of the Asian-native black hornet, *Vespa velutina nigrithorax,* were confirmed on Tsushima Island. Then, in 2017, the strongest and most harmful invasive insect, the South American-native fire ant, *Solenopsis invicta,* was discovered in Kobe Port, and it was reported that it had been landing in many port areas in Japan. All of these invasive insect species invade even human habitats and directly cause serious damage to humans, such as disrupting human dining tables or attacking humans with toxic stingers. So, it is necessary to prevent their spread by any means necessary.

Since these invasive species are also social insects, our research team applied the same principles used for the chemical control of the European bumblebees to control these invasive insect species. For the Argentine ant and the fire ant, we deployed meat-based bait containing insecticides, while for the Asian black hornet, we placed lactate drink stations containing the IGR insecticide in their distribution areas, in order to make the workers to carry the insecticides back to their respective nests.

These chemical controls have been effective, and the established populations of Argentine ants in the Kanto region has been almost eradicated. The control tests of Asian black hornets with pesticides has been proceeding smoothly on Tsushima Island, and it has entered the stage of comprehensive control. The invasion of fire ants into inland areas has been prevented so far.

The implementation of these chemical controls has been successful thanks to the understanding and cooperation of local governments and residents, and the coordinated efforts of stakeholders in the region have led to the success of the controls. Due to the fact that these invasive insect species directly harm humans, it is considered that stakeholder consensus was smoothly reached.

On the other hand, I believe that the technology and civic collaboration achievements gained from the project of controlling the European bumblebees were the important reason why we were able to quickly carry out the control of these dangerous invasive insects. Furthermore, I expect that analyzing these cases of dangerous invasive insect control can also provide feedback to promote the control of the European bumblebees.

## Further Reading

Dohzono I, Kunitake YK, Yokoyama J, Goka K (2008) Effects of an alien bumblebee on native plant reproduction through competitive interactions with native bumblebees. Ecology 89:3082–3092

Goka K (2010) Introduction to the special feature for ecological risk assessment of introduced bumblebees: status of the European bumblebee, *Bombus terrestris*, in Japan as a beneficial pollinator and an invasive alien species. Appl Entomol Zool 45(1):1–6

Goka K (2022) Conservation biology for the commercial insect trade in Japan: agricultural bumblebees and companion insects as examples. Sci Tech Rev 41:132–141

Goulson D (2010) Impacts of non-native bumblebees in Western Europe and North America. Appl Entomol Zool 45:7–12

Inoue MN, Yokoyama J (2010) Competition for flower resources and nest sites between *Bombus terrestris* (L.) and Japanese native bumblebees. Appl Entomol Zool 45:29–35

Kishi S, Goka K (2017) Review of the invasive yellow-legged hornet, *Vespa velutina nigrithorax* (Hymenoptera: Vespidae), in Japan and its possible chemical control. Appl Entomol Zool 52:361. https://doi.org/10.1007/s13355-017-0506-z

Sakamoto H, Goka K (2022) Efficiency of ant-control agents in colony-level oral toxicity tests using *Tetramorium tsushimae* (Hymenoptera: Formicidae) for post-establishment control of the red imported fire ant, *Solenopsis invicta* (Hymenoptera: Formicidae), vol 58, pp 25

Sakamoto Y, Kumagai NH, Goka K (2017) Declaration of local chemical eradication of the Argentine ant: Bayesian estimation with a multinomial-mixture model. Sci Rep 7:3389. https://doi.org/10.1038/s41598-017-03516-z

Tsuchida K, Kondo IN, Inoue NM, Goka K (2010) Reproductive disturbance risks to indigenous Japanese bumblebees from introduced *Bombus terrestris*. Appl Entomol Zool 45:49–58

Yokoyama J, Inoue NM (2010) Status of the invasion and range expansion of an introduced bumblebee, *Bombus terrestris* (L.), in Japan. Appl Entomol Zool 45:21–27

# Chapter 14
# Management of Wetland Ecosystems with the Participation of Citizens

**Jun Nishihiro**

**Abstract** Anthropogenic disturbances with moderate intensity and frequency are highly effective in conserving floodplain biodiversity. In addition to the approach of expanding activities aimed at biodiversity conservation, it is also important to create a situation in which activities with diverse "nonconservation objectives" result in biodiversity conservation. This chapter describes the retarding basins for flood control and abandoned rice fields as sites with the potential to serve as habitats for organisms that originally inhabited in flood plains. The activities for diverse purposes and their effects on biodiversity at these sites are described, and the conditions under which conservation can be achieved as a result are discussed. The existence of activities with common effects on biodiversity and mutually different objectives, i.e., redundancy of activities, would contribute to robust and long-term biodiversity conservation.

**Keywords** Floodplain · Retarding basins · Rice paddies · Anthropogenic disturbance

## 14.1 Introduction

Biodiversity decline is currently ongoing in various ecosystems worldwide, with freshwater ecosystems experiencing particularly notable losses (WWF 2022; https://livingplanet.panda.org/). In Japan, the loss of biodiversity in floodplain ecosystems is occurring at a significant rate (Washitani 2007). This can be attributed to the mountainous terrain that characterizes Japan, leading to the intense development of

J. Nishihiro (✉)
Center for Climate Change Adaptation, National Institute for Environmental Studies, Tsukuba, Ibaraki, Japan
e-mail: nishihiro.jun@nies.go.jp



Y. Suzuki-Ohno (ed.), *Community Science in Ecology*, Ecological Research Monographs, https://doi.org/10.1007/978-981-97-0304-3_14

urban and agricultural lands on lowland plains, most of which were originally river floodplains.

Human activities within floodplains, however, have not consistently resulted in a decline of biodiversity throughout history. In Japan, components of agricultural landscape such as rice paddies, reservoirs, and irrigation channels have historically served as essential habitats for organisms that originally inhabited river floodplains. This change can be viewed as a replacement by human activity with the moderate disturbance function once provided by river flooding (Fig. 14.1).

The rapid decline of floodplain biodiversity in Japan primarily occurred after the World War II. The mechanism can be divided into two pathways. The first is the expansion of urban and industrial development within floodplains. These lowland areas possess potential for floodplain ecosystems but are also favorable for development. Currently, approximately 70% of Japan's population resides on floodplains (Statistics Bureau of Japan 2003). Many rivers have undergone construction of large, uninterrupted embankments on both sides, leading to the loss of floodplains (Nishihiro et al. 2021a).

The second is that the modernization of paddy fields. With the spread of rice cultivation using large machinery, it became common practice to drain water completely from paddy fields for more than half of the year. This resulted in a marked increase in the efficiency of agricultural production, but also in a significant loss of the rice paddy's function of providing habitat for organisms that originally inhabited the floodplain (Katayama et al. 2015). Furthermore, increased use of herbicides and insecticides led to a further loss of biodiversity (Katayama et al. 2019).

Despite these trends, certain areas in Japan still possess potential for conserving floodplain biodiversity, such as retarding basins. Retarding basins, which temporarily store water during river floods, have the potential to provide habitats for floodplain organisms, as they are allowed to flood and are not subject to urbanization. There are at least 143 retarding basins in Japan (Suwa and Nishihiro 2020), and with the expected increase in heavy rainfall resulting from climate change, the role played by retarding basins in disaster-risk reduction is increasing. However, not all retarding basins possess high value as habitats of floodplain organisms. Moderate disturbance is necessary to maintain biodiversity.

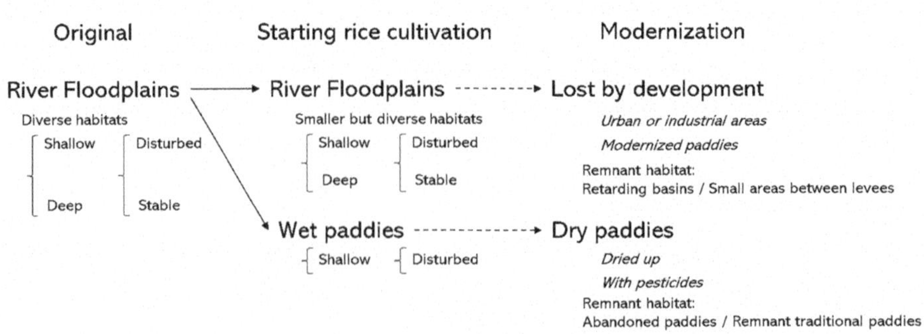


**Fig. 14.1** Overview of changes in floodplain habitats in Japan

Abandoned paddy fields represent another potential site for conserving wetland biodiversity. Currently, approximately 10% of all agricultural land in Japan is abandoned (Ministry of Agriculture, Forestry and Fisheries). If such fields have sufficient water table and experience moderate disturbance, they can contribute to the conservation of biodiversity (Koshida and Katayama 2018).

In the past, disturbances caused by river flooding or traditional agriculture have contributed to biodiversity conservation. Now that these activities have decreased, a new social system is necessary to continue moderate disturbance. Citizen activities in retarding basins and abandoned paddy fields could contribute to biodiversity conservation. Such citizen activities do not necessarily have to have biodiversity conservation as their main objective. Instead, a diversity of objectives may contribute to the sustainability of activities. This chapter explains citizen activities in wetlands in Japan and the significance of them in biodiversity conservation.

## 14.2 Citizen Activity in Retarding Basins

Retarding basins are riverine infrastructures that primarily serve to mitigate flood risks (Breen et al. 1994). Simultaneously, the space of retarding basins can offer a diverse range of functions, including the conservation of biodiversity and contributions to health, welfare, and education. In the Asahata Retarding Basin, Shizuoka city, Japan, public visitors and students, including those from special support schools, have participated in the utilization and management of the area (Nishihiro et al. 2021b). These various activities have significantly contributed to the conservation of the flora and fauna of the floodplain.

The Asahata Retarding Basin is situated in a space managed by the prefectural government and has been acquired from paddy fields for flood control purposes. During heavy rains, water overflowing from the river channel enters the retarding basin through overflow dikes. This reduces flood risk in the downstream area (Fig. 14.2). The space of the retarding basin is an invaluable green space in the highly urbanized city. A part of the basin is managed as a park by the city government.

Historically, the Asahata Retarding Basin area first transitioned from a floodplain swamp to a rice paddy field, and finally to the retarding basin. Reflecting this history, the flora and fauna that originally inhabited the floodplain, which are now in significant decline throughout the country, still exist in this area. The typical plant species are *Euryale ferox*, *Monochoria korsakowii*, and *Penthorum chinense*. These species germinate after soil disturbance. The presence of these plants is thought to have depended on the disturbance caused by river flooding when the site was a floodplain swamp, and on the disturbance caused by cultivation when it was a traditional paddy field. The frequency and intensity of flood-induced disturbance have decreased significantly since the area was developed as a retarding basin, and cultivation was ceased. Although floodwater enters the area once a year, the water stored in the retarding basin's space beyond the overflow bank has a slower velocity and is

**Fig. 14.2** The functioning Asahata retarding basin. (**a**) The river is on the right, and the retarding basin in on the left. Floodwaters overflowing the overflow levee. (**b**) Retarding basin during a flood

less likely to cause disturbance to the vegetation. For this reason, the areas of habitat of these plants have become small parts of the retarding basin.

Currently, citizen activities are the primary supporters of the habitat of floodplain plants and animals in the Asahata Retarding Basin. A park area set in the Asahata Retarding Basin was opened to the public in 2021. One of the park's features is utilization of natural wetland environment (Fig. 14.3). Since the park's opening, the author and staffs of the park management NPO have been collaborating with park visitors to create a wetland. The work referred to as "wetland creation" involves plowing the ground where no specific crops are grown, allowing children to play, and permitting diverse plants and animals to grow and inhabit the area (Fig. 14.3). In the management process, relatively large and dominant plant species, such as *Phragmites australis* and *Typha angustifolia*, are controlled, and sparse vegetation with shallow water depth is created. This activity is carried out monthly, with 10–20 participants each time. The activities are held on Sundays, and the participants are primarily parents who have their children participate in the environmental education program for primary school children held in the park.

In 2022, the 'wetland creation' activities were undertaken 7 times in a small wetland site measuring approximately 50 meters in length and 30 meters in width. A total of 19 aquatic insect species, three fish species, and 81 plant species were found at this site. Among the plants identified, six (*Tillaea aquatica*, *Penthorum chinense*, *Gratiola japonica*, *Veronica undulata*, *Monochoria korsakowii*, and *Chara braunii*) were the vulnerable or endangered species in Japanese national red list (Fig. 14.4). These species are small annual plants with disturbance-dependent characteristics. These plants are thought to have once inhabited river flood plains or traditionally farmed rice paddies and regenerated as results of anthropogenic disturbance by citizens. It should be noted that the wetland that was created through these activities displays unique characteristics when compared to other areas of the park. Children can engage in various recreational activities such as catching insects. When the participants of the 'wetland creation' program were surveyed, majority of them expressed positive feedback; 'activity using the sickle was joyful', 'the activity

**Fig. 14.3** Park in the Asahata Retarding Basin. (**a**) Suburban park area with playground equipment; (**b**) Large lawn. This area is also flooded during floods. (**c**) Wetlands imagined by citizens' activities; (**d**) citizens' 'wetland creation' activities; (**e**) Meeting for information sharing after activities

enhanced the personal health', and 'the organizers gave me a wonderful opportunity'. The activity of 'wetland creation', in which participants are given the freedom to utilize their creativity and physical prowess, is thought to be resulting a positive influence on their physical and mental well-being of those involved.

**Fig. 14.4** Endangered species which regenerated at the created wetland in Asahata retarding basin. (**a**) *Gratiola japonica*, (**b**) *Monochoria korsakowii*, (**c**) *Chara braunii*

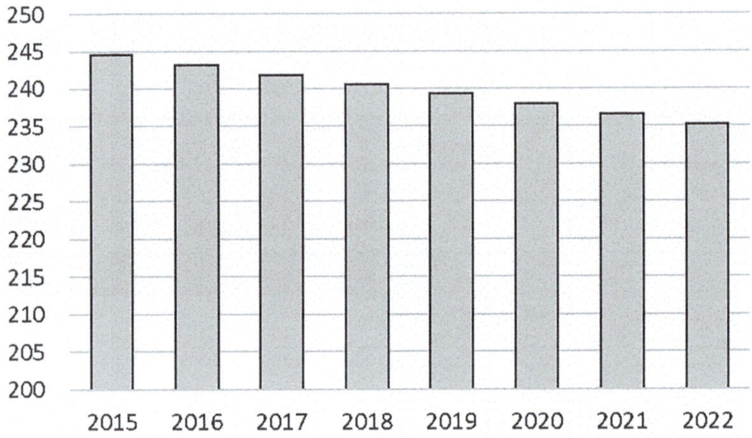


**Fig. 14.5** Area of cultivated paddy fields in Japan as a whole. Note that the minimum value in the graph is not zero. (Data source is from the Ministry of Agriculture, Forestry and Fisheries. https://www.maff.go.jp/j/tokei/sihyo/data/10.html)

## 14.3 Utilization of Abandoned Rice Paddies

Abandoned rice paddies are proliferating throughout Japan, largely due to a decrease in the number of successors to farmers (Fig. 14.5). Over time, abandoned paddies tend to become overgrown, and its infrastructure, such as irrigation and drainage channels are damaged, making it difficult to resume cultivation. In general, it is often thought that such abandoned paddy should be minimized and that the area should be developed into industrial land or other uses that contribute to economic productivity. However, abandoned rice paddies often have high value as natural capital. In general, lowland wetlands can provide flood risk reduction, water purification, and biodiversity conservation functions. These functions are also possessed by abandoned rice paddies. These ecological functions of wetlands are particularly high in lands that are difficult to drain or have high water table. Such paddy fields are more likely to be abandoned.

However, not all abandoned paddy fields possess high ecological functions. In some cases, a lack of water supply from springs or river floods makes wetland establishment difficult. Furthermore, even in locations where wetlands can be established from a topographical and hydrological perspective, if drainage facilities have been developed in the past when the land was used for agricultural purposes, the wetland may become increasingly dry. Nevertheless, appropriate management practices such as maintaining small banks for moderate water retention, mowing vegetation, and digging soil, can restore the biodiversity and ecosystem functions of wetlands. However, there are issues regarding who will implement ecosystem management, which was once carried out by farmers for food production, and for what purpose in the future.

In the watershed of Lake Inbanuma in Chiba Prefecture, citizens and some private companies are engaged in wetland ecosystem management activities in small valleys (Fig. 14.6). These small valleys, ranging from tens to hundreds of meters in length, are located on the edges of plateaus and are a common topographical feature in the Kanto region of Japan. There are approximately 500 such valleys in the Lake Inbanuma watershed. At the boundary between the valley slopes and the valley floor, groundwater frequently seeps out. In the past, rice cultivation was carried out using this spring water.

In a small valley in Tomisato City, Chiba Prefecture, Japan, locally known as Oyatsu, abandoned rice paddies in the valley floor and forests on the slope are being restored to improve their ecological functions. Oyatsu is a valley approximately 300 m in length surrounded by cultivated fields. Rice cultivation in the paddy fields ceased approximately 50 years ago. Although springs were present in the lower part

**Fig. 14.6** Topography of a small valley in Chiba prefecture, Japan

of the valley slope, the springs drained downstream without wetting the valley floor due to the existence of a concrete drainage channel between the slope and the valley floor surface. As a result, the valley floor has become increasingly dry, and high-density vegetation dominated by *Solidago altissima* and *Phyllostachys edulis* has developed, making it inaccessible.

In the summer of 2019, a project for wetland restoration was initiated at the site in question. In this activity, a group of local residents, mostly retirees, worked with a local NPO to remove overgrown vegetation and restore rice paddy banks and embankments (Fig. 14.7). The wetland restoration involved blocking a concrete drainage channel, diverting water to rice fields on the valley floor, and creating a shallow wetland. These efforts continued for nearly a year, and by the fall of 2020, the landscape of the valley floor had changed dramatically (Fig. 14.7).

The authors and collaborators are currently investigating the ecological functions and biodiversity of the wetlands that have been restored in the abandoned paddy fields. The concentration of nitrogen (nitrate ion) in the spring water was found to be excessively high, with levels exceeding 40 mg/L due to fertilizer application on the plateau. However, it was later confirmed that the concentration was reduced to

**Fig. 14.7** Wetland restoration in abandoned rice paddies in a small valley; (**a**) the field before the restoration; (**b**) citizens participated in the restoration; (**c**) restored wetland; (**d**) aquatic plant community established in the restored wetland

below 10 mg/L after passing through a regenerated wetland. Numerous aquatic plants have emerged on the former wetland surface. Whereas only 24 plant species had been recorded in the rice fields prior to the restoration, now 60 species were recorded, with the dominant species *Limnophila sessiliflora* and *Monochoria vaginalis*, and several endangered charophytes such as *Chara braunii*, *Nitella axilliformis*, and *Nitella megacarpa* var. *japonica* have also been observed (Fig. 14.7). These plants are thought to have germinated from propagules that once flourished during the period of paddy cultivation and remained in the soil.

Notably, the restoration of biodiversity and ecosystem functions is not the only outcome of this project. The local community, which initiated the natural regeneration, has established a new NGO to manage the site. Oyatsu now serves as a meeting place for people from the region and beyond. The main goal of the community members participating in the activities is to provide a playground for these children. The well-maintained wetlands at the bottom of the valley also house a vast number of firebugs, which drew in more than 300 visitors during the summer of 2021.

In the Lake Inbanuma watershed, abandoned paddy fields in small valleys are being restored and utilized in locations other than Tomisato City. The objectives of the projects vary based on the site, including biodiversity conservation, tourism use, and technological development for business, providing a place for children to experience nature, and rice cultivating in their own way. These projects contribute to the conservation of biodiversity and the improvement of ecosystem functions. Additionally, the "Satoyama Green Infrastructure Network" has been established as a network of diverse stakeholders to exchange information on the management and utilization of nature, and regular study meetings are held to exchange opinions.

## 14.4 Diverse Purposes and Multiple Functions

The biodiversity of floodplains had been maintained by moderate disturbance. However, the suppression of flood disturbances and the loss of paddy fields with habitat providing functions have made it necessary to implement anthropogenic disturbances in wetlands for effective biodiversity conservation. Determining the appropriate level of disturbance is crucial for conservation. Too weak disturbance is ineffective, while too strong disturbance can have irreversible effects on topographic and hydrological features.

Activities that cause disturbances do not necessarily have to be aimed at biodiversity conservation, as different objectives can also have positive effects on biodiversity. For example, in the Asahata retarding basin, various activities with mutually different objectives, such as harvesting plants for traditional crafts and creating paddy fields for educating the disabled, took place and had similar effects on biodiversity conservation (Nishihiro et al. 2021b). Similarly, in the small valleys of Chiba Prefecture, various activities with different objectives, including nature-education activities, small-scale organic farming, and technological development for companies, contributed to the growth of disturbance-dependent endangered species.

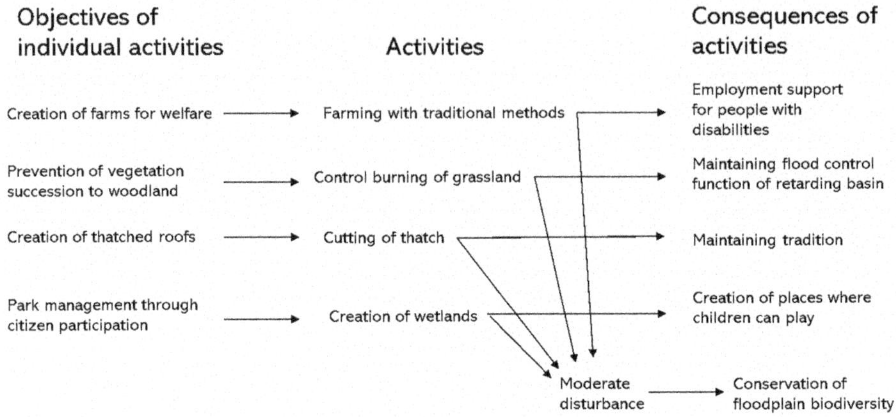


**Fig. 14.8** Relationship between objectives recognized in the activities, activities actually carried out, and their results. Activities with mutually distinct objectives can result in moderate disturbance and contribute to biodiversity conservation. An example is given for Asahata Retarding Basin

Studies on ecosystem stability have proposed that ecosystems with community redundancy, i.e., a diversity of species within groups that share the same ecosystem function, are more resilient to external perturbation (Naeem 1998; Biggs et al. 2020). In other words, response diversity within functional groups contributes to the stability of ecosystems. A similar phenomenon can be observed in ecosystem management. It can be expected that the existence of various activities that have similar impacts on biodiversity will enhance the sustainability of ecosystem management. Activities with sole purpose may not be sustainable. For example, the sustainability of an activity whose sole purpose is education will be affected by fluctuations in the number of children in the area, while an activity whose sole purpose is economic return will be heavily influenced by market fluctuations. However, the coexistence of activities with different objectives can be expected to produce management practices that are resilient to both social and environmental change (Fig. 14.8).

To ensure diversity or redundancy of activities, it is important to make a system that fosters the involvement of stakeholders with diverse backgrounds. In the Asahata retarding basin, a council under the Law on the Promotion of Nature Restoration are playing an important role. The council is open to all stakeholders interested in the conservation and utilization of nature in the area. When it was first established, the Nature Restoration Council of the Asahata retarding basin consisted mainly of groups and individuals whose primary objective was biodiversity conservation. However, in 2014, the council strengthened its links with activities aimed at welfare, such as supporting people with disabilities, and a more diverse group of stakeholders, including private companies and NGOs, began to participate. In the Lake Inbanuma basin, the Satoyama Green Infrastructure Network includes citizen groups and individuals involved in ecosystem management activities with a variety of objectives (~120 registered members on the mailing list as of May 2023). The network holds monthly study meetings where information is shared and discussed.

A web-based GIS is also being developed to share the content of citizens' activities and the scientific knowledge on which they are based. The existence of such networks, which facilitate the participation of diverse actors, provides opportunities for cooperation between activities with different mutual objectives.

Scientific evidence on the impact of ecosystem management would be helpful in expanding activities with diverse objectives. Information on the state of biodiversity is expected to bring new perspectives to organizations that have embarked on activities with objectives other than biodiversity conservation. Increased scientific knowledge and its dissemination to the public will contribute to an increase in the number of actors involved. The contribution of research not only on ecosystems but also on human dimensions would be significant. Many studies suggest that outdoor activities have positive effects on human physical and mental health. Sharing these results to society at large is considered to enable the involvement of a more wide range of stakeholders.

## References

Biggs CR et al (2020) Does functional redundancy affect ecological stability and resilience? A review and meta-analysis. Ecosphere 11:e03184

Breen PF, Mag V, Seymour BS (1994) The combination of a flood-retarding basin and a wetland to manage the impact of urban runoff. Water Sci Technol 29:103–109

Katayama N, Baba YG, Kusumoto Y, Tanaka K (2015) A review of post-war changes in rice farming and biodiversity in Japan. Agric Syst 132:73–84

Katayama N et al (2019) Organic farming and associated management practices benefit multiple wildlife taxa: a large-scale field study in rice paddy landscapes. J Appl Ecol 56:1970–1981

Koshida C, Katayama N (2018) Meta-analysis of the effects of rice-field abandonment on biodiversity in Japan. Conserv Biol 32:1392–1402

Naeem S (1998) Species redundancy and ecosystem reliability. Conserv Biol 12:39–45

Nishihiro J, Taki K, Harada M, Miyazaki Y, Kawaguchi Y, Miyashita T (2021a) Dynamics of society and ecosystem 5. History and future of Japanese rivers. Asakura-shoten, Tokyo. (in Japanese)

Nishihiro J, Koike S, Ono A (2021b) Biodiversity conservation through various citizen activities in a flood control basin. In: Green infrastructure: function, implementation and governance. Springer

Statistics Bureau of Japan (2003). https://www.stat.go.jp/english/data/tanshin/5.html

Suwa Y, Nishihiro J (2020) Distribution and location characteristics of flood-control reservoir in Japan. Ecol Civ Eng 23:85–97. (in Japanese with English summary)

Washitani I (2007) Restoration of biologically-diverse floodplain wetlands including paddy fields. Glob Environ Res 11:135

WWF (2022) Living planet report. https://livingplanet.panda.org/

# Chapter 15
# Editor's Note

**Yukari Suzuki-Ohno**

**Abstract** Ecological community science (citizen science) programs can contribute to scientific research, environmental education, artistic and social activities, alien species elimination, and conservation. Resolving the shortage of human resources, funds, and facilities for ecological community science is essential for the future development of ecological community science programs. The shortage of human resources and facilities is closely related to the shortage of funds. The reasons behind fund shortages in Japan include the cutting of management expense grants for national universities, the declining number of students due to the declining birthrate, and the high maintenance costs of museums for their buildings and specimens. Several ecological community science programs described in the previous chapters resolved this shortage by incorporating the programs into environmental education in schools or by relying on intermediary support organizations. Ecologists need to firmly convey the value of ecological community science and increase the number of supporters for ecologists. For the future development of ecological community science programs in Japan, it is also essential to increase the consideration of the diversity of participants in the programs. The diversity of participants in an ecological community science program sometimes creates gaps between the ecologists who plan the program and the participants in the program. Ecologists need to consider participants' diversity and avoid the negative effects of the gaps. Collaborating with researchers in the fields of education, psychology, sociology, and informatics will produce fruitful results in ecological community science programs.

**Keywords** Community science · Citizen science · Environmental education · Intermediary support organization · Participants' diversity

Y. Suzuki-Ohno (✉)
Tohoku University, Sendai, Miyagi, Japan



Y. Suzuki-Ohno (ed.), *Community Science in Ecology*, Ecological Research Monographs, https://doi.org/10.1007/978-981-97-0304-3_15

## 15.1 Looking Back on the Previous Chapters

The previous chapters demonstrated that ecological community science programs have various aims, target organisms, methods, and outputs. The aims and target organisms of ecological community science programs generally determine the methods used in the programs and the methods affect the types of participants and outputs in the programs. Ecological community science programs can contribute to scientific research (Chaps. 3, 4, 5, 6, 8, and 9), environmental education (Chaps. 3 and 6), artistic and social activities (Chaps. 5 and 7), alien species elimination (Chaps. 9 and 13), and conservation (Chap. 14). For the future development of ecological community science programs in Japan, it is essential to resolve the shortage of human resources, funds, and facilities for ecological community science (Chaps. 3, 4, 5, and 8) and to increase the consideration of the diversity of participants in the programs (Chaps. 3, 4, 11, and 13). Several ecological community science programs described in the previous chapters resolved and increased them. I would like to address the keys to resolving and increasing them in the following sections.

## 15.2 Human Resources, Funds, and Facilities for Ecological Community Science

Few human resources can engage in ecological community science in Japan, even though ecological community science is one of the effective means of ecological research, environmental education, and conservation activities. A shortage of human resources is becoming apparent in ecology and various other fields in universities, research institutes, and museums in Japan. In universities and research institutes, most scientists are too busy to contribute to community science, environmental education, and conservation activities. In addition, these contributions do not improve the scientist's evaluation and are conducted only by the goodwill and unpaid efforts of scientists. Even in museums and science centers, most curators are too busy to contribute to community science, given their miscellaneous tasks and a shortage of curators. Curators working in community science and environmental education are often overworked. The shortage of human resources and facilities is closely related to the shortage of funds in Japan. In the case of Japanese universities, management expense grants for national universities have been reduced by more than 12% when comparing 2004 and 2020 (Ministry of Education, Culture, Sports, Science and Technology 2020). It is becoming increasingly difficult to obtain budgets due to the cutting of management expense grants and the declining number of students due to the declining birthrate. In the case of Japanese museums, the average budget per museum in the 2010s has decreased by about one-third compared to the 1990s because the number of museums increased in the 1990s, but the total budget for museums from local public entities has gradually decreased (Ministry of Education, Culture, Sports, Science and Technology 2023). While most museums have a small budget, they must pay high maintenance costs for their buildings and specimens. Even the National Museum of Nature and Science, the most famous museum in Japan, started crowdfunding to deal with the steep rise in energy bills in 2023. Additionally, museum facilities are expected to play a role as tourism spots (Japan Tourism Agency 2018), and some museums will shift to tourism rather than environmental education and ecological community science. Before people engaging in ecological community science disappear in Japan, it is necessary to improve the evaluation of ecological community science and environmental education, improve the evaluation of people who contributed to them, and increase the number of human resources, funds, and facilities for them.

### *15.2.1 Ecological Community Science and Environmental Education in Schools*

The evaluation of ecological community science and the number of people contributing to ecological community science may increase if ecological community science programs contribute toward environmental education in primary, junior, and high schools. Teachers in junior and high schools in Japan face problems such as shortages of teaching materials and programs, difficulty in curriculum management, and a lack of cooperation with external organizations for environmental education (Ministry of Environments 2021). Teachers will be able to conduct classes for environmental education by cooperating with ecological community science programs. However, teachers rarely have direct contact with the ecologists who plan them. Even if they were connected, the workload may be too heavy for teachers, or the research objectives of the ecologists may differ from the educational goals of the schools. Ecologists who plan ecological community science programs need to find ways to make the teachers' tasks easier (Chap. 3) or find suitable schools to participate in the program (Chap. 6).

### *15.2.2 Intermediary Support Organizations for Community Science*

Ecologists who plan the programs can try and resolve these issues themselves. However, intermediary support organizations could help coordinate ecologists and schools (Chap. 6) and consult with ecologists to find suitable schools to participate in their programs or ways to make the workload for teachers easier. Intermediate support organizations may also be able to resolve the funding problem because they can act as a bridge between ecologists and corporations. The international situation of Corporate Social Responsibility (CSR), Sustainable Development Goals (SDGs) activities, Environment, Social, and Governance (ESG) investment, and Taskforce on Nature-related Financial Disclosures (TNFD) (https://tnfd.global/) has increased the demand for corporations to demonstrate consideration for biodiversity. Target 15 in the Kunming-Montreal Global Biodiversity Framework requires corporations to regularly monitor, assess, and transparently disclose the risks, dependencies, and impacts on biodiversity (https://www.cbd.int/gbf/targets/15/). In response to these issues, corporations expected the ecological community science programs to play a significant role. If an ecologist cooperates with a corporation for an ecological community science program, the ecologist may obtain funding but be unable to use or publish collected data as they belong to the corporation. However, if corporations provide funding to intermediary support organizations who support ecologists in organizing the program and collecting data, the ecologists do not need to ask corporations for permission to use their data or for funding. In Japan, some NGO/NPO organizations such as Earthwatch Japan (https://www.earthwatch.jp/), general incorporated associations such as the Sustainability Center (https://m-sustainable.org/), and companies established by scientists such as Think Nature (https://think-nature.jp/en) and Biome (https://biome.co.jp/en/) can be intermediary support organizations. It is ideal to increase the number of such intermediary support organizations for environmental education.

Increasing the number of intermediary support organizations and the amount of funding for ecological community science depends on whether people know their necessity and importance. Ecologists need to firmly convey the value of ecological community science and ecology itself to increase the number of various supporters for ecologists. This will help resolve the shortages in human resources, funds, and facilities. In addition, it is necessary not to reduce the number of intermediary support organizations for ecological community science. When intermediate support organizations collaborate with ecologists who plan ecological community science programs, the ecologists must avoid imposing the work solely on the intermediate support organizations and work in partnership with the intermediate support organizations and participants.

## 15.3 Diversity of Participants in Ecological Community Science Programs

The depth of ecological community science allows for diverse volunteers in the programs. The diverse participants mean that there are many differences in knowledge, skills, and motivation. The differences in knowledge and skills have the potential to obtain accurate as well as inaccurate data (Dickinson et al. 2010). However, inaccurate data often result from a difficult survey method (e.g., species identification by only nonexperts) or complex explanations (e.g., use of jargon) in data collection. A simple survey method and sufficient explanation about data collection will reduce the risk of obtaining inaccurate data (Chaps. 3 and 4).

The motivation difference may cause a difference in the benefits provided by the program and those expected by participants. In such cases, some participants may drop out of the program. Needless to say, the program must ensure the freedom of participants, including their choice of nonparticipation. However, considering motivation differences will lead to sustainable ecological community science programs.

### *15.3.1 Motivations of Volunteers in Ecological Community Science*

In this section, I would like to consider the motivation of volunteers reported in previous studies. With the success of online community sciences such as Foldit (Cooper et al. 2010) and Galaxy Zoo (Clery 2011), scientists have focused on the motivation of the volunteers participating in them. Interviews with volunteers suggested that their motivation shifted from initial egoistic ones to altruistic and collective ones after a long-term engagement (Rotman et al. 2012). The initial egoistic motivation was the volunteers' interest in the target, methodology, and experiences of the community science program. They thought that the program satisfied their intellectual curiosity or helped build their careers. After a long-term engagement in the program, they felt that they belonged to the community of the program and had altruistic and collective motivations for the community.

As there are arguments regarding the distinction between egotism and altruism, Clary and Snyder (1999) and West and Pateman (2016) classified the motivations of volunteers into six types: "Understanding," "Values," "Social," "Enhancement," "Protective," and "Career" (Table 15.1). Participants in ecological community science programs mainly possess these six motivations (West and Pateman 2016). The questionnaire survey in the fish survey project using eDNA showed that the major motivation of participants is the motivation to acquire new knowledge (86.1%), which corresponds to the "Understanding" motivation (Suzuki-Ohno et al. 2023). Simultaneously, many participants had motivations to contribute to science (80.6%) or protect the environment (47.2%), which corresponds to the "Values" motivation. In addition, a certain number of participants had motivations associated with volunteering activity, which can be classified into the "Social," "Enhancement," "Protective," and "Career" motivations. The questionnaire survey for participants of Monitoring Sites 1000 (Chap. 7) demonstrated that the attachment to their monitoring site had the most influence on their motivation to participate in conservation activities. The attachment to their monitoring site would correspond to the "Values" motivation. For participants with less than 5 years of experience, sympathy for the activity and enjoying activities with others were strong motivations to continue the survey. These correspond to the "Social" motivation.

**Table 15.1** Classification of the motivations of volunteers discussed in previous studies (West and Pateman 2016)

| Function | Example |
|---|---|
| Understanding | Wanting to learn new things |
| | Wanting to share existing knowledge with others |
| Values | Helping other people |
| | Helping the environment |
| | Helping a particular site |
| | Helping science (i.e., contributing to scientific research) |
| Social | Wanting to increase social relationships (e.g., meeting new people) |
| Enhancement | Wanting to improve themselves personally through volunteering |
| Protective | Wanting to reduce negative feelings or to address personal problems |
| Career | Hoping to gain experience that will benefit their future careers |

### *15.3.2 Targeting and Information Sharing*

Diversity in participants' motivation sometimes creates a motivation gap between ecologists and participants. For example, if an ecologist is only interested in academic research and starts a program, participants with high "Social" motivation may be less likely to feel a benefit from the program. Such a motivation gap can be partially resolved by setting the appropriate aims for ecological community science programs. Setting the appropriate aims and message of the program (i.e., targeting) will lead to the acquisition of volunteers with that motivation, automatically increasing their satisfaction and continuity.

Even if ecologists have appropriately set the aim of the program, there are some cases where the satisfaction of participants decreases as the information they have requested is not provided. The aim of contributing to science will attract participants who want to get new knowledge, but their satisfaction may be low if the results and progress of surveys are not shared frequently. The aim of contributing to environmental education will attract teachers who want their students to do independent research projects in schools, but their satisfaction may be low if the data collected in the program cannot be used for their research projects immediately. The aim of contributing to conservation will attract participants who want to help the environment, but their satisfaction may be low if they do not know how their work contributes to conservation. Ecologists who plan ecological community science programs should be careful about information sharing that matches the participants' motivations. Mutually honest and respectful communication will help participants understand the ecologists' ideas. In some cases, face-to-face communication will be effective because some participants may have the "Social" motivations.

## 15.4 Conclusion

Ecological community science programs in Japan face the problem of a lack of human resources, funds, and facilities. In some cases, this problem was overcome through collaboration with schools and intermediate support organizations. As social demand for ecological community science and environmental education is increasing, ecologists need to firmly convey the value of ecological community science to resolve this shortage.

Ecological community science programs often show the diversity of participants in the programs. Ecologists must consider the diversity of participants and be careful about equity in ecological community science programs. It may seem unrelated to ecology, but community science is originally a cross-disciplinary subject. If possible, collaborating with researchers in the fields of education, psychology, sociology, and informatics, and with community science programs in other fields, will produce fruitful results. If a universal design of community science programs is realized, it can be applied to many programs and should lead to the prevalence of ecological community science. The prevalence of ecological community science will lead to real-time big-data collection and classification and will support the further development of ecology. This will also lead to effective environmental education and conservation activities, breaking the barrier between experts and nonexperts and making ecology a long-lived discipline that benefits many people.

Even without planning an ecological community science program, ecologists can get involved in ecological community science in various ways, such as cooperating to identify species and organizing, analyzing, and publishing data. Organizing, analyzing, and publishing the data of ecological community science programs can help preserve information regarding today's biodiversity for future ecologists and provide solutions to the unsolvable problems of ecology and conservation.

## References


Clary EG, Snyder M (1999) The motivations to volunteer: theoretical and practical considerations. Curr Dir Psychol Sci 8:156–159

Clery D (2011) Galaxy evolution. Galaxy zoo volunteers share pain and glory of research. Science 333:173–175. https://doi.org/10.1126/science.333.6039.173

Cooper S et al (2010) Predicting protein structures with a multiplayer online game. Nature 466(7307):756–760. https://doi.org/10.1038/nature09304

Dickinson JL, Zuckerberg B, Bonter DN (2010) Citizen science as an ecological research tool: challenges and benefits. Annu Rev Ecol Evol Syst 41:149–172

Japan Tourism Agency (2018). https://www.mlit.go.jp/kankocho/news02_000354.html (in Japanese)

Ministry of Education, Culture, Sports, Science and Technology (2020). https://www.mext.go.jp/content/20201104-mxt_hojinka-000010818_4.pdf (in Japanese)

Ministry of Education, Culture, Sports, Science and Technology (2023). https://www.mext.go.jp/b_menu/toukei/001/index05.htm (in Japanese)

Ministry of Environments (2021). https://www.env.go.jp/content/900497869.pdf (in Japanese)

Rotman D et al (2012) Dynamic changes in motivation in collaborative citizen-science projects. In: Proceedings of the ACM 2012 conference on computer supported cooperative work. https://doi.org/10.1145/2145204.2145238
Suzuki-Ohno Y et al (2023) Evaluation of community science monitoring with environmental DNA for marine fish species: "fish survey project using environmental DNA". Environ DNA 5:613–623. https://doi.org/10.1002/edn3.425
West S, Pateman R (2016) Recruiting and retaining participants in citizen science: what can be learned from the volunteering literature? Citiz Sci 1:1–10. https://doi.org/10.5334/cstp.8

The manufacturer's authorised representative in the EU is Springer Nature Customer Service Centre GmbH, Europaplatz 3, 69115 Heidelberg, Germany. If you have any concerns regarding our products, please contact ProductSafety@springernature.com

Printed and bound by CPI Group (UK) Ltd, Croydon, CR0 4YY
30/09/2026
02225253-0001